Fractalize That!

A Visual Essay on Statistical Geometry

That!

Fractals and Dynamics in Mathematics, Science, and the Arts: Theory and Applications

ISSN: 2382-6320

Published

Forthcoming

Fractals and Dynamics in Mathematics, Science, and the Arts:  Vol. 3 Theory and Applications

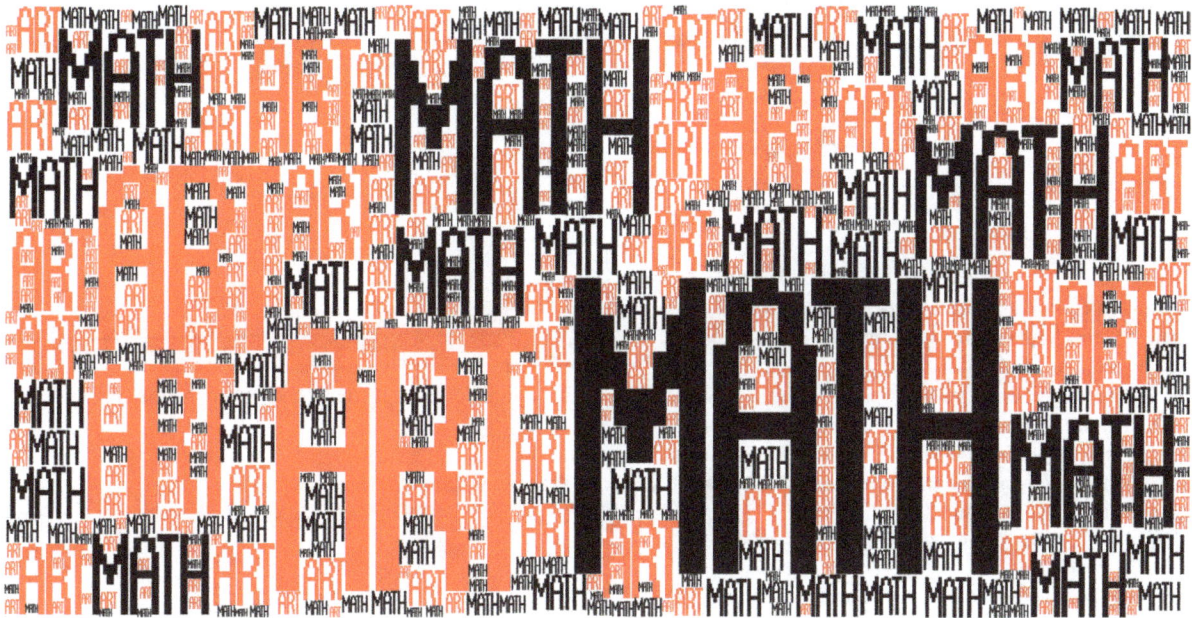

Fractalize That!

A Visual Essay on Statistical Geometry

John Shier

 World Scientific

NEW JERSEY · LONDON · SINGAPORE · BEIJING · SHANGHAI · HONG KONG · TAIPEI · CHENNAI · TOKYO

Published by

World Scientific Publishing Co. Pte. Ltd.

5 Toh Tuck Link, Singapore 596224

USA office: 27 Warren Street, Suite 401-402, Hackensack, NJ 07601

UK office: 57 Shelton Street, Covent Garden, London WC2H 9HE

Library of Congress Cataloging-in-Publication Data

Names: Shier, John (Writer on geometry), author.

Title: Fractalize That! a visual essay on statistical geometry / by John Shier.

Description: New Jersey : World Scientific, 2019. | Series: Fractals and dynamics in mathematics, science, and the arts. Theory and applications ; volume 3

Identifiers: LCCN 2018044019 | ISBN 9789813275164 (hc : alk. paper)

Subjects: LCSH: Tessellations (Mathematics) | Fractals. | Algebraic spaces. | Geometry, Algebraic.

Classification: LCC QA166.8 .S55 2019 | DDC 516/.132--dc23

LC record available at https://lccn.loc.gov/2018044019

British Library Cataloguing-in-Publication Data

A catalogue record for this book is available from the British Library.

For any available supplementary material, please visit
https://www.worldscientific.com/worldscibooks/10.1142/11126#t=suppl

Desk Editors: V. Vishnu Mohan/Kwong Lai Fun

Typeset by Stallion Press
Email: enquiries@stallionpress.com

To my father

George Raymond Shier

A man who inspired an interest in science and
mathematics in his sons from an early age.

Preface

What is offered here is a new way of completely and randomly filling space with ever-smaller non-touching fill regions. Any fill region shape can "tile the plane" by this technique. The patterns are good examples of random fractal construction, and show fractal behavior in a visual way which is easy to grasp.

The title reflects the invention of a new verb. As it emerged from ongoing work in which no fill region shape was found for which this technique does not work, the word *fractalize* was coined. It is convenient for planning and describing new images.

Because the images are generated by a computer using quite simple but flexible mathematical rules, it can be viewed as a book about mathematics and computer graphics. There is also value here for physicists, graphic designers, architects, and builders of computer models. In another sense, it can be thought of as an art book since many of the patterns are attractive for decoration.

There is a temptation to think that computer-generated images are rigid and mechanical, but the present work differs from much previous work in the use of random numbers, which imparts an interesting variety. Much is left to the inventiveness and skill of the person creating software for these images. The evidence shows that any fill region shape can be randomly "fractalized" by this technique. The key feature of the images is a perception of *receding to the infinitesimal*. Large fill regions are placed, and then continually smaller ones in a steady downward progression, constantly filling in the ever-tinier spaces between the earlier fill regions. Available evidence shows that this process can be continued "to infinity". It does not halt when properly set up.

Physicists are offered a new way of looking at space in several Euclidean dimensions. Any finite region of space can be thought of as fractally filled in a vast number of ways.

The mathematician is offered a new space-filling construction whose properties are quite engaging and sometimes surprising. It has been explored little by formal mathematics.

The general reader is offered some varied and interesting images, and visual examples of fractals which may help in understanding them. Since the 1977 book by Mandelbrot, fractals have been found in many areas of science and technology, and an understanding of them can be important in several fields. The book's images present a blend of order and randomness, and some will see connections with the space-filling geometric art of Escher.

The process here described and illustrated is defined in purely mathematical terms. Where does it fit within mathematics? It does not fit comfortably within an existing category. Fill region shape is a quite important aspect, and shape is traditionally the province of geometry. The non-integer power law and the Hurwitz zeta function belong to calculus. Statistics is inextricably involved through the use of random search. This is not mathematical packing because none of the fill region shapes are mutually touching. The thoughtful reader will see many loose ends which could be the starting point for further studies. Because the author has needed a label for his work, he has adopted the name "statistical geometry" in his writing. Paul Bourke's name is "random tiling" which expresses a different aspect of the algorithm.

The smaller fill regions are quite tiny in some of the images, and the printing process does not always fully resolve them.

Chapter 1 on space-filling patterns sets the stage by reviewing what is known and introducing the new technique without formal mathematics.

Chapters 2–5 take up the fractalization of various fill region shapes from traditional geometry. The manner in which the fill regions fit together is not "all the same" but is much affected by the details of the shape, so that numerous examples are needed to show the variety of filling patterns. Along the way the effects of the parameters c and N are pointed out, and such phenomena as order–disorder and correlation are treated descriptively.

Chapter 6 is devoted to statistical geometry examples, and shows the great variety and flexibility which is possible.

Chapters 7–9 are for the mathematically trained, and go into the details of the algorithms and describe their properties.

Chapter 10 contains advice and examples for those who want to "do it myself".

The origin of our subject lies in recreational computation. The author was amusing himself by creating computer-generated art, and found one of these patterns by accident. Benoit Mandelbrot was also important. The connection is that Mandelbrot wrote a book and I read it. Discovery comes to the prepared mind. Had I not read the book I would never have grasped that I was looking at a fractal pattern.

For brevity, the interesting subject of three-dimensional statistical geometry is omitted, but the reader is assured that the algorithm works quite well for that case. The only three-dimensional example is the image of fractalized toroidal rings on the cover, which was created by Paul Bourke.

About the Author

John Shier has a BS from Caltech and a PhD from the University of Illinois, both in physics. He was involved in the integrated circuits industry for 30 years, starting with 5 μm features and ending at 300 nm features. The last 20 years of this work involved low-noise analog integrated circuits, which impressed him with the randomness which exists in the physical world. Following retirement from ICs, he taught introductory physics and electrical engineering courses at Normandale Community College in Bloomington, MN for several years. The accidental discovery described in the book took place in 2010 while doing recreational computation.

Acknowledgments

I would like to thank Paul Bourke for several contributions cited in the book. Paul has been an enthusiastic colleague since the summer of 2011 and the exploration of the subject has benefited greatly from his work and the mutual sharing of results. Paul was the first to look into the use of additional random variables besides x, y (such as the rotation seen in Fig. 2.8 of Chapter 2) and did all of the early work on three-dimensional fractals. Prof. Steve Butler is thanked for supplying the data used to construct Fig. 1.4 in Chapter 1. Andy Webster and Ed Jensen helped answer my C questions. I would also like to thank Julie Johnson, Chris Innes, Doug Dunham, Clint Sprott, Nat Friedman, Barry Cipra, Dietrich Stoyan, and Cye Waldman for their encouragement and enthusiasm.

John Shier
Apple Valley, MN

October, 2018

All things flow from random chance or human will.

Contents

Chapter 1

Introduction: Space-Filling Patterns

Ways of filling space with tiles have been known since ancient times. If we have square or rectangular tiles it is easy to fill any spatial area by simply placing more and more of them side-by-side. Millions of bathrooms and shower stalls have tiled walls of this kind. With a particular area to be filled and a particular tile size, only a finite number of tiles are needed to cover the area. There are many other ways to cover an area with tiles having the shapes of triangles, hexagons, etc. The formal word for a tile pattern is *tessellation*.

Figure 1.1 shows several simple tessellations. If all of the tiles have the same color, we lose all sense of separate pieces, so it is traditional to use different colors in a regular sequence. The checkerboard of Fig. 1.1(a) is one of the most common examples; it uses two colors. Another common example is shown in Fig. 1.1(b). It uses a diamond having 60° and 120° corner angles and requires three colors. The diamonds have three orientations which are rotated 120° from each other. One must color the three orientations differently in order to see the pattern, and with

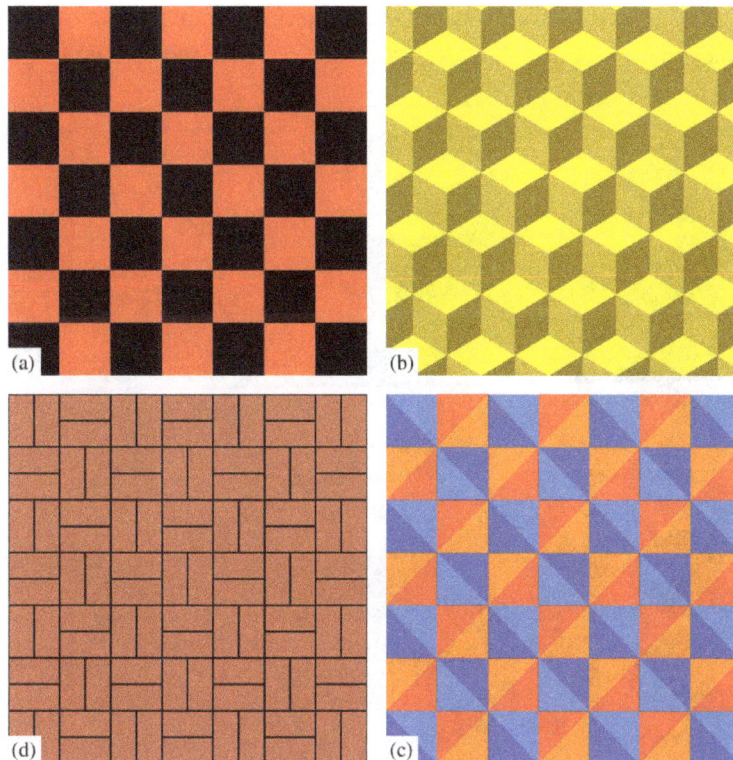

Fig. 1.1. Tessellations. (a) A checkerboard formed from squares. (b) "Tumbling blocks" made by tiling 60°–120° diamonds. (c) A tiling using 45° triangles; an example of a Truchet tiling. (d) A pattern sometimes seen in brick sidewalks.

1

a suitable choice of shading the image has a strong three-dimensional illusion — the eye perceives polyhedral bumps sticking out of the plane.

Figure 1.1(c) shows a tessellation made with symmetric triangles having two 45° corners. Two such triangles joined in a square is called a "Truchet tile". There are four possible orientations for the triangles, shown by the four colors used. A huge number of repeating patterns can be made with these tiles, and mathematical studies of the possible arrangements have been done. The paired rectangles pattern shown in Fig. 1.1(d) is often seen in brick sidewalks, and only works if the bricks have a width half their length.

It can be shown mathematically that there are 17 possible symmetries [1] for planar tessellations where every piece repeats periodically as we move left–right and up–down as in Fig. 1.1. In the 1970s, the mathematical physicist Roger Penrose showed that it is also possible to take two simple polygonal shapes and fit them together to fill all space in a manner which does not repeat: a non-periodic tessellation.

Quite interesting uses for tessellations were invented by the Dutch artist Maurits Escher in the late 20th century [2]. Escher's idea was to distort the sides of the tiles in such a way that we retain the basic space-filling property, while teasing the shape into that of a familiar object such as a fish or bird. Figure 1.2 shows an example of a fish tessellation with a twist. At the left, the fishes are simple black and white repeating shapes. As we move to the right, the fishes gradually acquire unique colors and become distorted into shapes which are no longer the same, while they remain space-filling. The fishes at the right seem to be trying to swim out of the picture.

Tessellations fill a finite region of the plane with a *finite* number of shapes. In the early 20th century, geometric constructions were invented which fill a finite region with an *infinite* number of regular shapes. One of the best known of these is the Sierpinski [3] triangles.

The construction of Fig. 1.3 begins with an empty black triangle. An oppositely-pointing blue triangle is placed in the middle of it, creating three new smaller empty triangles in the corners. We then proceed to place new, smaller green triangles in these empty spaces, which create yet more empty triangles, which in turn get yet smaller light green triangles placed, etc. The procedure follows a regular pattern of "generations" or "recursions", with each new generation having three times more triangles than the previous one, each with a side length half that of the previous

Fig. 1.2.   A tessellation constructed with a fish shape. The regularly repeating symmetric black–white shapes at the left progressively shift to distorted and colored shapes at the right.

Fig. 1.3.   The 6th generation of a Sierpinski fractal construction which fills a black triangle with an infinite number of smaller triangles.

generation. Figure 1.3 shows six generations, each with a different color, and it can be seen that the remaining unfilled black area is already quite small.

Those familiar with Sierpinski's work will see that I have reversed the usual procedure in which a filled triangle is successively emptied by cutting out smaller triangles. The "cutting out" method is used in mathematics, while the "adding" method described here has been adopted by physicists in studies of how material particles pack together [4–6]. The additive approach is much more attractive for visual art. Where usage differs we will follow the physics conventions.

The Sierpinski triangles never quite fill the whole region; there is always some area left over even after a huge number of generations. We call the left-over region the *gasket*. It is easy to see that even after a large number of generations, the total *area* of all the inserted triangles remains finite, with a limiting value equal to the area of the initial triangle. If, however, we consider the total *perimeter* of the triangles, it grows *without limit*. This paradoxical behavior is characteristic of *fractals* [3]. The construction of a fractal involves ever-finer fracturing or fragmentation of the region which is to be filled. This fragmentation property of a fractal can be described mathematically in terms of a *fractal dimension D* which is not an integer [3].

Mathematicians make a distinction between a disc and a circle. A disc contains all of the points within a circular boundary. A circle contains only the points on the curved line. We follow this convention in the book.

Most people will look at a circle and say "There is no way to completely fill a planar region with discs. They don't fit together smoothly." But if we are willing to consider the use of an *infinite* number of discs, it is possible to make constructions which are space-filling "in the limit" where we allow an infinite number of ever-smaller discs. These patterns have received the name "Apollonian circles" and are of several kinds depending on what bounding region we set out to fill. Figure 1.4 shows an example.

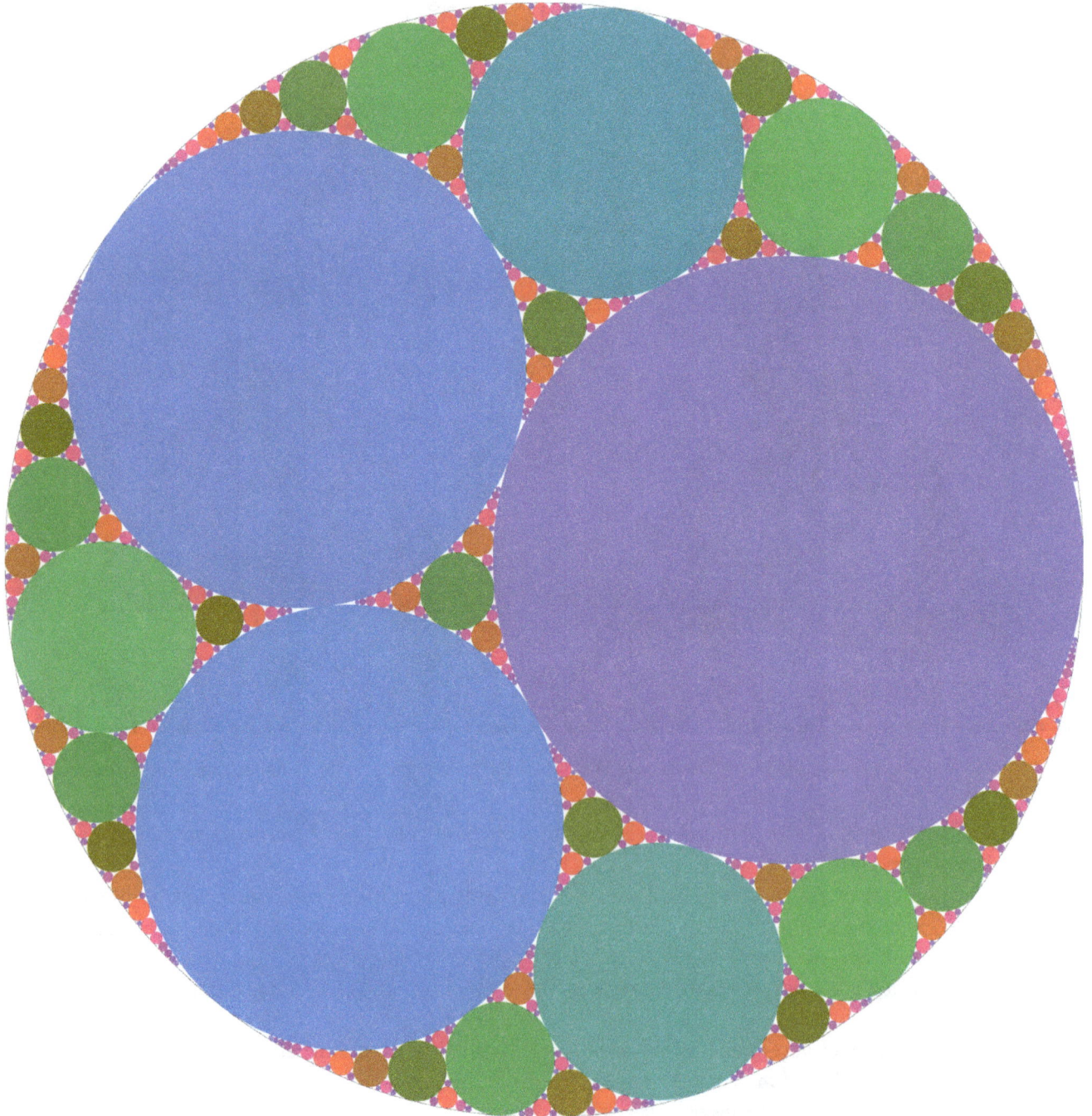

Fig. 1.4.   A touching-discs construction which fills a bounding circle "in the limit" with an infinite number of small discs. It has been given the name "Apollonian circles" after the ancient Greek mathematician Apollonius who showed that given any three circles, it is possible to find a fourth circle which is tangent to all of them. It makes an interesting comparison with Fig. 1.6 and Fig. 4.1 in Chapter 4. Log-periodic color. Data courtesy of Steve Butler.

The circles are placed by taking existing circles three at a time and constructing smaller circles which are tangent to all three. Like the Sierpinski triangles, the Apollonian circles have a fractal dimension D which is not an integer [3].

In the fractal space-filling patterns described so far, every fill region has an exact size and position. The fill region shapes are the simple shapes of elementary geometry. In what follows, we will see that if we are prepared to abandon this regularity, it is possible to create space-filling *random* fractals [7, 8] where the sizes of the fill regions follow a regular rule but their positions do not. The surprise is that it is apparently possible to do this for any fill region shape, including quite odd "blobby" ones unknown to any geometric text.

Before taking up the images generated by this new technique, it is desirable that the reader understand something about how they are created. Without some insight into the process, it is difficult to grasp what is going on. The first step is to assume a spatial region to be filled, which has an area A. It is usually a square or rectangle in the examples here, but in principle it can be any shape. It is to be filled with ever-smaller fill regions, and the next step is to generate the sequence of areas for these regions according to a mathematical rule called a *power law*. In a particular example, the sequence might run 1.6870, 1.1277, 0.8851, ... and keep on getting smaller. These numbers can be thought of as areas in square inches or other convenient units. We omit details, but the sum of these areas (to infinity) is just the total area A which is to be filled. Many such area sequences are possible, with the details depending upon the choice of two numbers used in generating them — the *power-law exponent* c and the *starting point* N. A mathematical account of the algorithm can be found in Chapter 7.

After finding the area sequence, we choose a particular fill region shape (such as a circle, square, or crescent). We proceed with the areas in order of size, beginning with the largest. For the largest area we find the fill region's

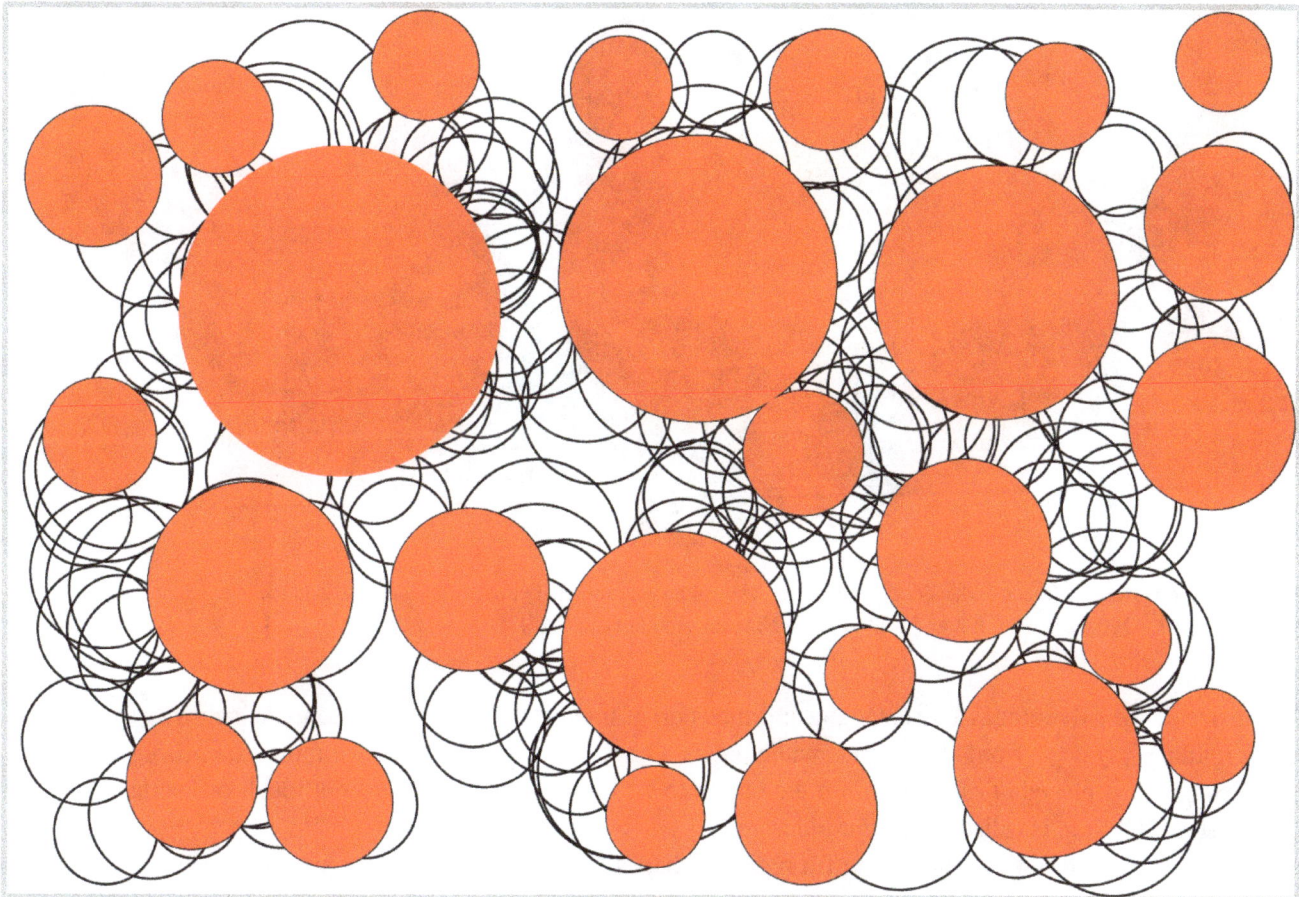

Fig. 1.5.   An illustration of the process for discs in a rectangle. Each black-line circle shows a random trial. After 25 discs were placed, a red disc was drawn at the location of each *successful* trial (placement). It is evident that there were many unsuccessful overlapping trials.

Fig. 1.6.   A random disc fractal with a circular boundary. The colors are random, but done in a way that excludes very dark and light colors. $c = 1.49$, $N = 3$, 1200 discs, 95% fill, and white gasket. Note the similarity to Fig. 1.4.

linear size. If the fill region is a disc, the linear dimension of interest will be the i-th radius, which we can find from the formula $r_i = \sqrt{A_i / \pi}$ from elementary mathematics. Other fill region shapes have more complicated formulas. The first shape is placed *at random* within the region to be filled, such that it does not touch or overlap any of its boundaries. This step is called the *initial placement*.

We then proceed with the following fill regions (shapes) in order of decreasing size. For each fill region, we find its linear dimension(s) and then try it *at a random place* within the region to be filled. This random step is called a *trial*, and at this random location the shape may overlap with some previously placed fill region or with a boundary, in which case the trial fails and we do another trial with the same fill region at a different *random* location. We keep on repeating the random trials until we succeed in finding a location where the fill region does not overlap or touch any of the previously placed fill regions or the boundary. This is called a *placement*. We enter the linear dimension(s) and the location (x, y coordinates) in a list of placed fill regions and go on to the next-smaller fill region. Thus the process proceeds (see Fig. 1.5):

0. First fill region, initial placement at random, write placement data in the list.
1. Next fill region, trial, trial, trial, ... success after n_{t1} trials ... write placement data in the list.
2. Next fill region, trial, trial, trial, ... success after n_{t2} trials ... write placement data in the list.
...

The areas of the discs must obey a precise rule given in Chapter 7 which ensures that they are "space filling in the limit".

Figure 1.6 shows how the process works in practice for the particular case of discs randomly fractalized within a circular boundary. The discs are rather tightly packed here. While there is at least a tiny space between adjacent discs, they don't have much "wiggle room". For the smallest discs, the average number of trials needed for placement was about 50,000. The radius of the smallest disc is only about 1.2% of the radius of the biggest one.

The reader can see that it has taken only a modest number of placed discs to reach a state where the bounding circle is quite full. This is typical of cases where the parameter c is near its upper limit. The orderly generation-by-generation structure of Sierpinski is gone here, but this structure is also a fractal. Close study of the mutual arrangements of nearby discs between this image and the Apollonian circles of Fig. 1.4 shows some similarities.

As with most infinite processes, we find that "full" is a mirage which (like infinity itself) recedes the closer we get. We can achieve 90%, 99%, ... fill, but we can reach 100% only in the idealized world of calculus where one nonchalantly "passes to the limit". In the world of images drawn on paper, we must make our choice of "How full is full enough?"

The following chapters describe the filling arrangements of various fill region shapes in more detail.

Chapter 2

Squares and Rectangles

The reader will see parameters c and N mentioned frequently in Chapters 2–5. Precise mathematical definitions for them can be found in Chapter 7. We provide graphical information on what these parameters do without the use of formal mathematics. Figure 2.1 shows how the fill fraction varies with c for two different N values.

- With $N = 1$ (upper graph) when $c = 1.4$, the first fill region takes up about 33% of the bounding region. With decreasing c, the first fill region is much smaller.
- The bounding region fills much more rapidly with higher c in all cases. This has the consequence that the fill regions are much closer together for higher c values. This effect is quite visible in several of the images in this chapter.
- Filling is much slower when $N = 4$ than when $N = 1$. This is a general feature of higher N.

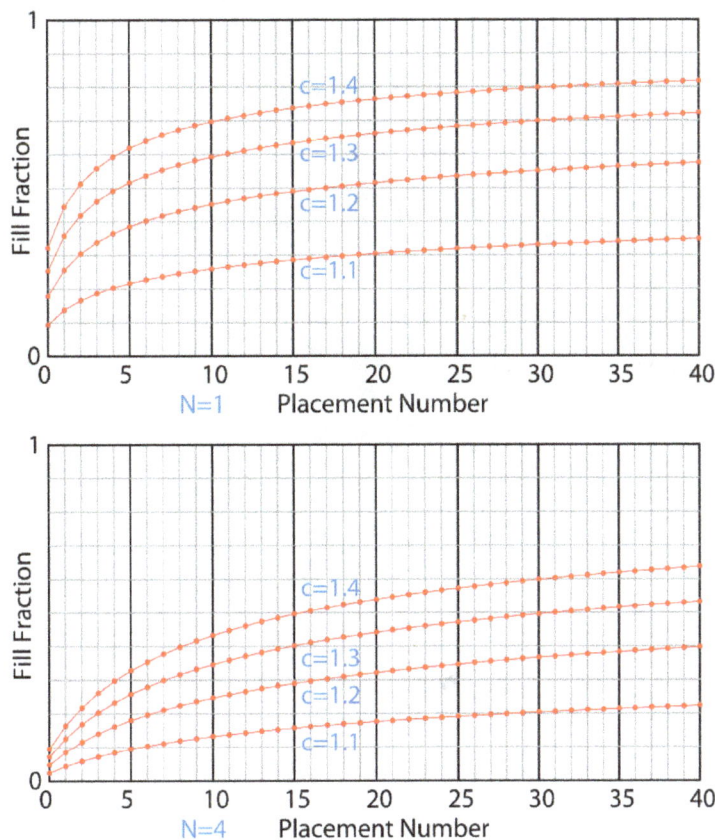

Fig. 2.1.   Fill fraction versus the number of placed fill regions for two N values. The dots indicate that the fill fraction is only defined for integer values of placements.

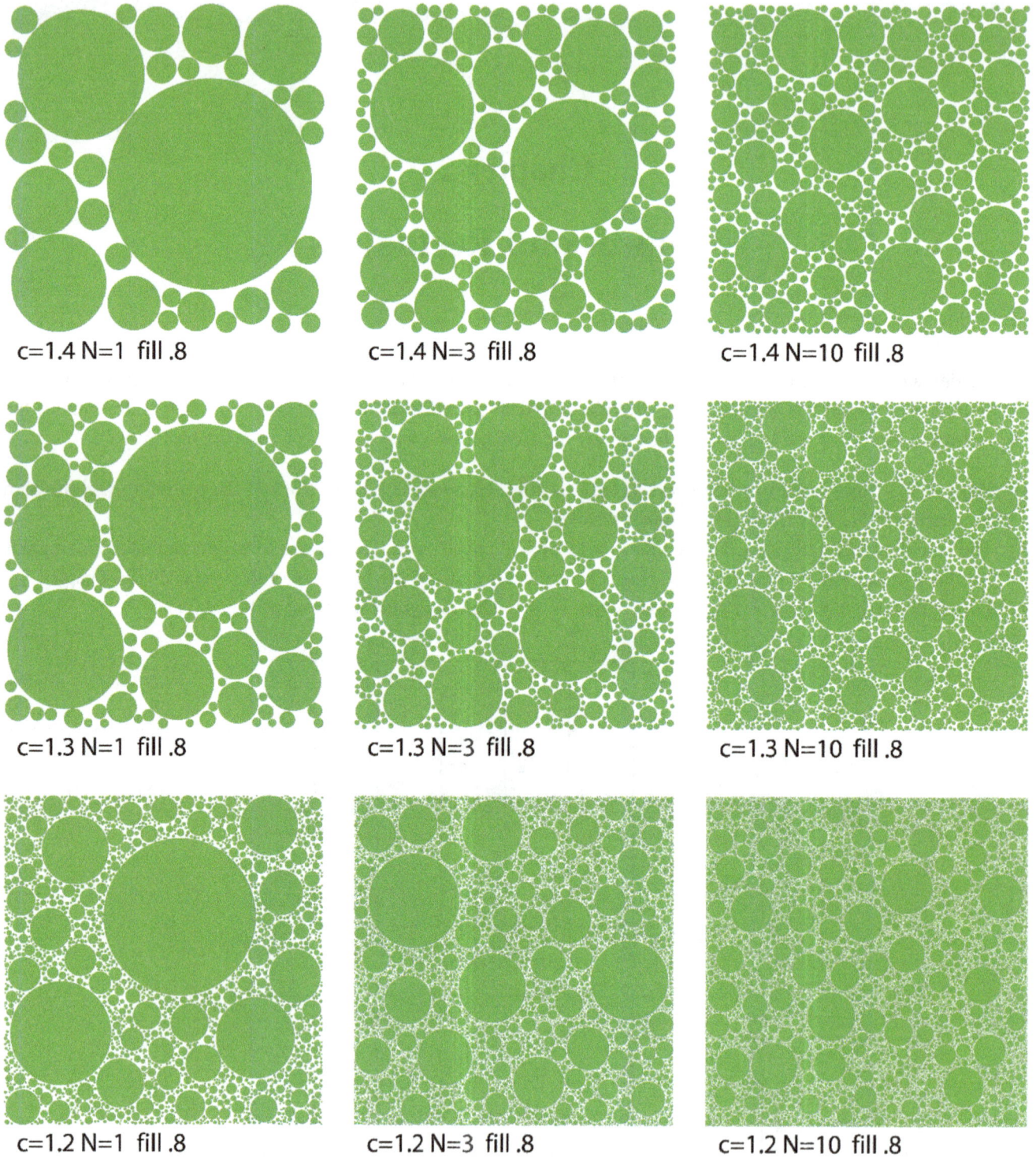

c=1.4 N=1 fill .8   c=1.4 N=3 fill .8   c=1.4 N=10 fill .8

c=1.3 N=1 fill .8   c=1.3 N=3 fill .8   c=1.3 N=10 fill .8

c=1.2 N=1 fill .8   c=1.2 N=3 fill .8   c=1.2 N=10 fill .8

Fig. 2.2.   Fill patterns for discs in a square with a fixed fill fraction. The number of discs needed to achieve 80% fill varies hugely. At the upper left, 32 discs suffice for 80% fill. At the lower right, it needs 29,690 discs. It can be seen that as one moves from left to right along a row, each pattern is a scaled-down randomized copy of the one to its left.

Figure 2.2 shows disc fill patterns for several values of c and N.

The reader who understands the patterns in Fig. 2.2 will see reflections of them in the images in Chapters 2–6. Or the images can simply be viewed as abstract art. Similar behavior of squares can be seen in Fig. 2.5.

Squares are a good place to start because they are a simple shape and have the shortest code for the algorithm. The code for checking overlap is simpler than for any other fill region shape and the range of c values which can be used is the widest: $1 < c < {\sim}1.49$ when $N = 1$, and up to $c \sim 1.56$ for higher values of N. It is possible to achieve higher percentage fills for squares than any other shape within the constraints of practical run times. Figure 2.3 shows an example. The gasket is white, but the eye has a hard time picking it out when it is only 3% of the area.

Fig. 2.3. Randomly fractalized squares. Log-periodic color, white gasket. In this color scheme fill region shapes of the same size have the same color. $c = 1.50$, $N = 3$, 3000 squares, and 97% fill.

Color schemes make a large difference in how the patterns are perceived. One might think that black on white would be ideal, but in fact the eye tends to see the smaller fill regions as a gray blur, especially with a high fill factor. Several color schemes are used in the book.

In a checkerboard, the squares alternate in color so that adjacent squares have opposite color. This helps greatly in seeing the pattern. The analog of a checkerboard for a random fractal is shown in Fig. 2.4(a) where the color alternates black–white–black … as we proceed starting with the largest square. It is thus a kind of randomized fractal checkerboard. The red gasket provides good visual contrast with both black and white squares. Figure 2.4(b), shows log-periodic color. Most of the properties of the fill regions (such as their areas and linear dimensions) give a straight line when plotted on a log–log plot against placement number. It is thus natural to use a color scheme with the same logarithmic variation. The color is made to vary *periodically* with the logarithm of linear dimensions, so that the color sequence repeats for ever-smaller fill regions. In Fig. 2.4(b), the smallest features are just starting to repeat the purple of the largest ones. One can also get good contrast by using random colors as shown in Fig. 2.4(c). The colors in Fig. 2.4(c) don't include all colors but are selected randomly from a lighter region of RGB color space.

Figure 2.5 shows what happens with squares when we vary the power-law exponent c while keeping the percentage fill the same. Figures 2.5(a)–(d) go clockwise from the upper left. This allows us to have images with two adjacent c values physically adjacent for easier comparisons. The most important effect is that as c falls, there is a huge increase in the number of placements needed to reach a given fill factor. In Fig. 2.5, the squares change color in proportion with the logarithm of their size (log-periodic color). In Fig. 2.5(c), the colors cycle through a complete period and back to red again, while in Fig. 2.5(d), they go through approximately two complete color cycles. The c value is highest for the upper left image, and falls progressively as we move clockwise.

In the previous examples, the squares always have the same orientation. It is possible to randomly rotate the squares with the result seen in Fig. 2.6. The manner in which the rotation angle of the squares is varied must be understood. The angle is assigned a random value between 0° and 90° at *each trial*, like the x, y position coordinates. Thus the angle is not fixed but has the value at which the first random trial was successful. The c value is fairly high in Fig. 2.6 so that the squares have little "wiggle room". (The effect of c on "wiggle room" is best seen in Fig. 2.9 where c is varied with a fixed number of fill regions.) As a result, the first successful trial for a given placement tends to have an orientation angle much the same as adjacent previously, placed squares. Random orientation angles of this kind were first explored by Paul Bourke.

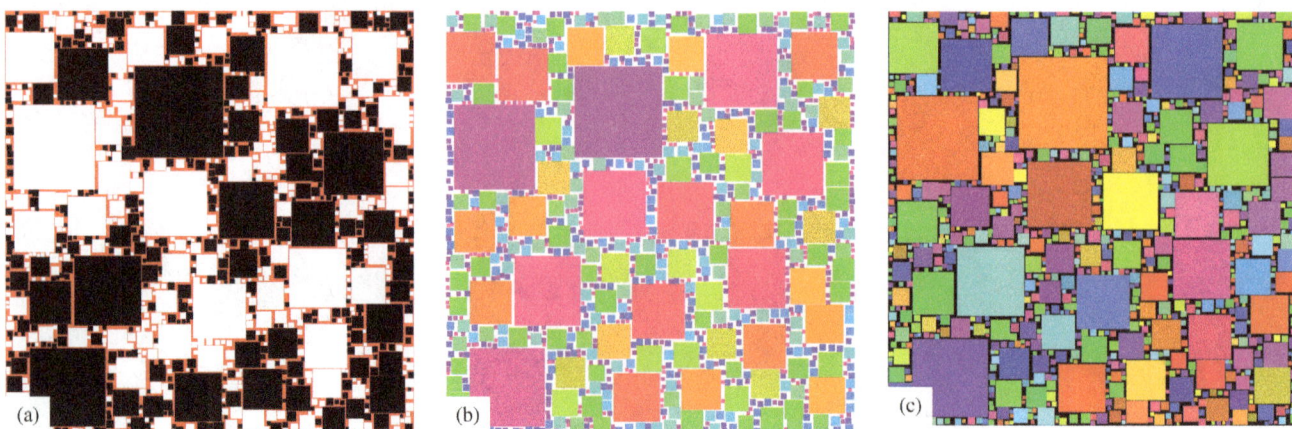

Fig. 2.4.   Color schemes. The set of squares is the same in each image. (a) "Devil's checkerboard" color scheme. Red gasket. (b) Log-periodic color, with squares the same size having the same color, white gasket. (c) Random color with a black gasket. $c = 1.38$, $N = 8$, 1000 fill squares, and 84% fill.

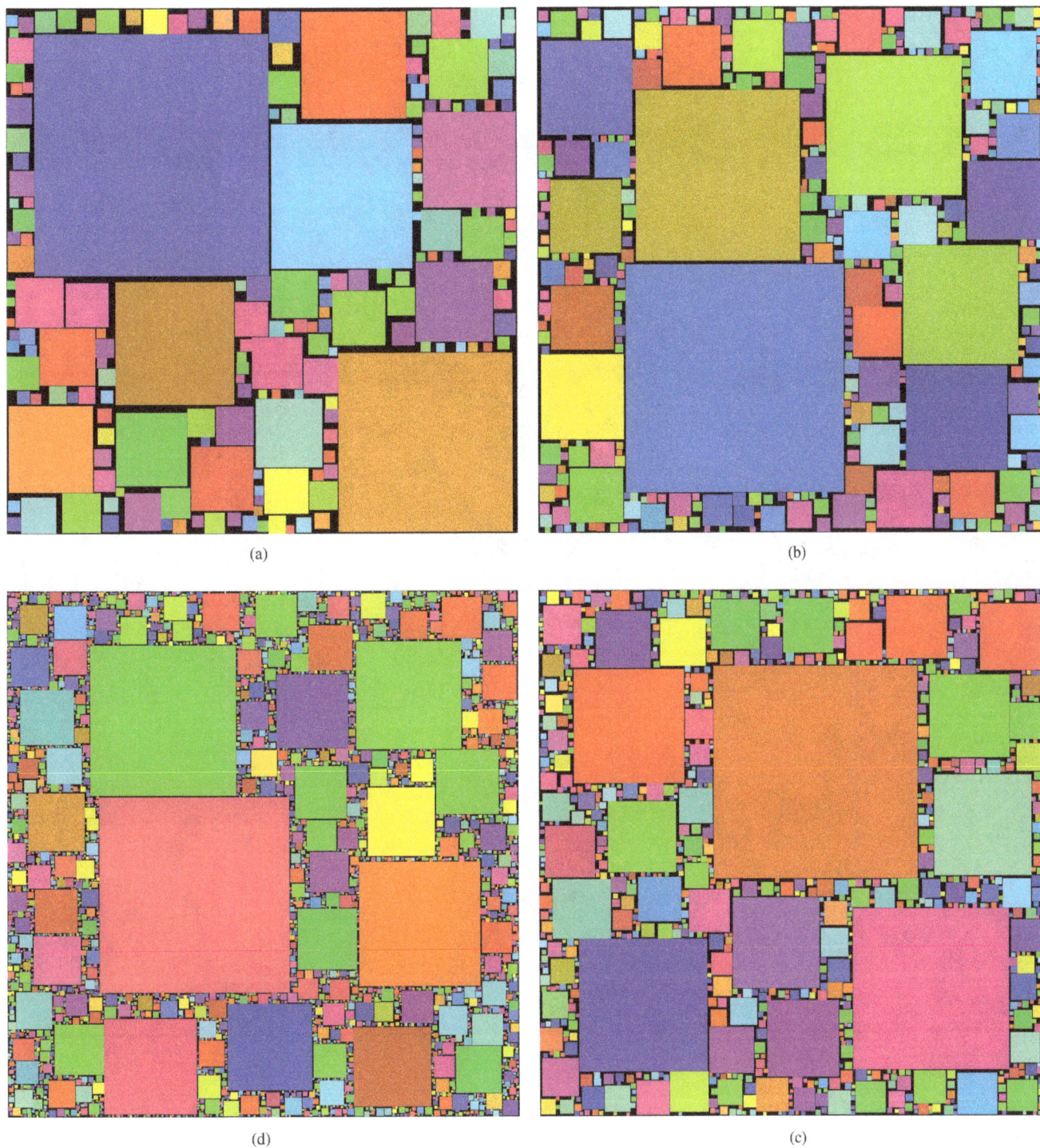

Fig. 2.5. The effect of the exponent c with the same percentage fill (88%). (a)–(d) clockwise from upper left. (a) $c = 1.48$, and 151 squares. (b) $c = 1.40$, and 381 squares. (c) $c = 1.32$, and 1517 squares. (d) $c = 1.24$, and 15,118 squares. In all cases, $N = 2$. Random color, black gasket.

Fig. 2.6.   Fractalization of squares with a random orientation angle which varies at each trial. The color is a function of the orientation angle, with the same color indicating the same orientation angle. Periodic boundaries, black gasket. $c = 1.43$, $N = 3$, 2500 squares, and 95% fill. The smallest fill regions required an average of 30,000 trials per placement.

In Fig. 2.6, the color is varied cyclically according to the rotation angle, and goes through a complete cycle in 90°. It can be seen that there are "islands" of similar colors. They arise because the few first-placed large squares strongly influence the orientation of later ones which are placed nearby.

Another feature of Fig. 2.6 is *periodic boundaries*. A fill region is permitted to fall slightly outside of the image space, and if it intersects the left boundary, the same shape is also placed in the list at a position $x + L_x$ where L_x is the width of the image in the x direction, and similarly for y. Thus, several pieces of the fill region are present

Fig. 2.7.   Two adjacent random-angle square fractal "tiles" with periodic boundaries. Each feature in the left half repeats in the right half. Log-periodic color, black gasket. $c = 1.43$, $N = 3$, 500 squares each tile, and 90% fill.

in the finished image. With an image made in this way, a larger repetitive image can be created by "tiling" several images together.

Figure 2.7 shows an example of an image with periodic boundaries tiled horizontally. It has log-periodic color (color by size) and one can see the smallest squares "adjusting" their orientation to the nearby larger ones, thus providing an example of order arising from a random process.

A rectangle can be thought of as a stretched square. The ratio of length to width is called the *aspect ratio* and provides another variable like the rotation angle which is internal to the fill region. Figure 2.8 shows what happens when we fractalize rectangles with alternating 3:1 and 1:3 aspect ratios. The white rectangles are elongated vertically, while the black ones are elongated horizontally. The choice of which fill region shape to use alternates at each placement. Thus the number of 1:3 and 3:1 rectangles is forced to be the same. For a given placement, all of the trials use the same aspect ratio.

The process runs smoothly for rectangles, although it does require more trials on average than simple squares. What is noteworthy is that there is quite strong correlation in the positions of the vertical (white) and horizontal (black) rectangles. The image consists of mostly white and black regions, with small islands of the opposite color at all length scales. The boundaries have the same effect as a nearby rectangle; the rectangles adjacent to the horizontal boundaries are largely black, while the vertical boundaries have white neighbors.

From a computational viewpoint, the two rectangle aspect ratios are thought of as *two distinct fill region shapes* and must be accounted for separately. In this book, two fill regions are only thought of as having the same shape if one can be made to coincide with the other by a uniform expansion or contraction. If a rotation or uniaxial stretch is required to bring them into coincidence, we view them as different fill region shapes.

The random trials in the process tell us that it is random, but the results often show a striking order, as in Figs. 2.6 and 2.8. How does this order arise from disorder? The agent for order is the *constraint* arising from non-overlap and the power law. Each new fill region has limited "wiggle room" and must conform to the previously placed fill regions if the trials are to find a place for it. It is this "feedback" from previous fill regions that imposes

Fig. 2.8.   Two rectangular fill regions fractalized with 1:3 and 3:1 aspect ratios alternating at placement. Red gasket. $c = 1.30$, $N = 5$, 3300 rectangles, and 86% fill.

order. Larger c values impose tighter filling for a given number of fill regions and the imposed order depends strongly on c with the largest ordering effects seen for c near its upper limit (Fig. 2.9). With quite low c values (slightly above 1), the patterns in the fill regions are quite random.

Figure 2.9 shows how the parameter c affects "degree of order". In all cases, 1200 rectangles were placed, with $N = 5$. For the highest c value Fig. 2.9(a) one can see the same pattern of "white near white and black near black". It can also be seen that in Fig. 2.9(a) the places available for further placement of fill regions having the size of the smallest ones shown are quite few and constricted, and that "vertical" holes mostly occur near vertical rectangles and

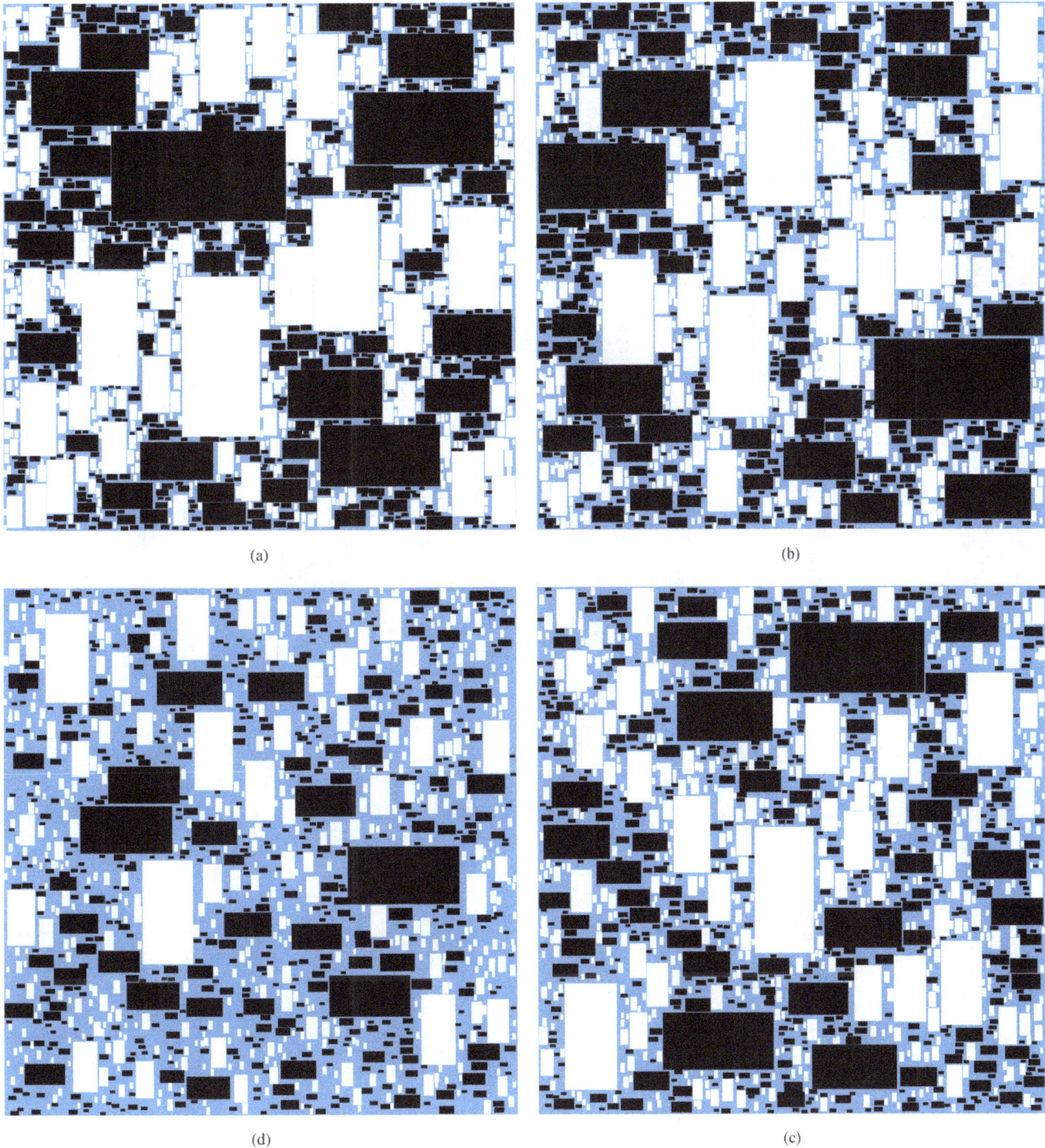

Fig. 2.9. Order and disorder; the effect of c. Two rectangular fill region shapes are fractalized with alternating aspect ratios of 2:1 and 1:2. $N = 5$, 1200 rectangles. Blue gasket. Figures (a)–(d) run clockwise from the upper left. (a) $c = 1.30$, 51,356 trials, and 82% fill; (b) $c = 1.24$, 20,347 trials, and 74% fill; (c) $c = 1.18$, 9033 trials, and 63% fill; (d) $c = 1.12$, 4475 trials, and 49% fill.

vice versa. For the lowest c value (Fig. 2.9(d)) the correlation pattern is largely absent with black and white rectangles thoroughly intermixed. The places available for further placements in Fig. 2.9(d) are quite abundant and impose few limitations on their orientation. This is also reflected in the approximate tenfold difference in the number of random trials needed to place this fixed number of fill regions. There is a steady drop in degree of order from Figs. 2.9(a)–2.9(d).

Fig. 2.10.   The words EARTH, AIR, FIRE, and WATER sequentially fractalized using letters made up of rectangles. $c = 1.11$, $N = 1.5$, 800 fill region shapes (200 of each word), and 52% fill. From 14 to 27 rectangles per word.

This qualitative pattern is seen for all fill region shapes and at all length scales. We see that c plays the role of an "order parameter" which allows one to arbitrarily choose how much order or disorder the pattern is to have.

From a computational viewpoint squares and rectangles have the fastest overlap test of any planar fill region shape. This means that the overlap test for any fill region composed of several rectangles is simple to implement. One tests all of the rectangles in fill region A against all of the rectangles in fill region B, quit on first fail. This might seem to cause too-long computation times, but in practice up to ~50 rectangles in a fill region shape is feasible. An interesting application of such rectangle-generated fill regions is *word art* — fractalization of fill regions which are words. Individual setup code can be created for each of the 26 Roman letters, so that space-filling patterns with words of up to 8–10 letters are possible with only a few minutes of setup work. Figure 2.10 shows an example. Some will recall that the Greek philosopher Aristotle claimed that earth, air, fire, and water were all the elements needed to make up the world.

The constraint of all rectangles with sides along the x and y axes produces a rather odd-looking font, but it seems unlikely that any previous font designer has operated under this limitation. It is often desirable to make the letters tall and narrow; a too-long word may not fit within a reasonable boundary.

Chapter 3

Triangles and Diamonds

Triangles with a single orientation do not fill very well. Figure 3.1(a) is near the upper limit of c for equilateral triangles. At high c, the triangle points always come quite close to neighboring triangles; there is little "wiggle room". As c decreases, "wiggle room" expands. Four of the triangles have been colored red to point out to the reader that there is a *motif* in the pattern for high c values. The central red triangle has three smaller ones as nearest neighbors. Close study of the rest of the pattern shows that this arrangement persists everywhere, especially for the smaller triangles. Comparison with Fig. 1.3 in Chapter 1 shows that this is the Sierpinski motif, somewhat randomized. In Fig. 3.1, as for the rectangles of Fig. 2.5 in Chapter 2, we find the correlation and order fading as c decreases.

An important feature of Fig. 3.1(a) is that most of the gasket regions where more triangles can be placed ("holes") are roughly triangular but the holes have a different symmetry. If the black triangles are thought of as "up arrows", most of the white holes are "down arrows".

The common patterns between randomly fractalized triangles and the completely ordered Sierpinksi fractal (Fig. 1.3 of Chapter 1) are clearer in Fig. 3.2 where the boundary is a triangle. The four-triangle motif shown in red in Fig. 3.1(a) is even more prevalent here. There are several forms of correlation in this image and close study of it is rewarding.

The holes which have opposite symmetry to the placed triangles offer a possibility for fractalizing alternating triangles pointing "up" and "down", as seen in Fig. 3.3. We have the exact opposite situation from the rectangles of Fig. 2.8 in Chapter 2. There the black and white rectangles were strongly clustered together. Here we have the opposite; the near neighbors of black triangles are almost all white triangles and vice versa. The two types of triangle are thus strongly *anticorrelated*. The maximum c value is much higher here than for the single-orientation case of Fig. 3.1.

Figure 3.4 shows four 45° triangles. They are placed in sequence with each type of triangle placed at every fourth placement, so that there are equal numbers of each type. Referring back to Chapter 1, we see that these are a randomization of the same type of triangle which makes nice tessellations. (Fig. 1.1(c) in Chapter 1 shows one of many possible examples.) The blue and aqua triangles both have sides which slope down to the right, while the red and orange triangles both have sides sloping up to the right. The image has very strong clustering and correlation of several kinds. Blue and aqua triangles are highly clustered and are mutually anticorrelated. Red and orange triangles are highly clustered and are mutually anticorrelated. This can be understood by looking at the holes available for new placements. In the blue-aqua regions, the holes almost all favor blue-aqua placements, for example.

When we fractalize three diamonds, we get the pattern seen in Fig. 3.5. One can see that each diamond type likes to "stick together". The red diamonds, for example, are concentrated in just a few regions. These are the same three diamonds which make the "tumbling block" tessellation (see Fig. 1.1(b) of Chapter 1). The clustering can be understood by a close study of the available holes. Only the few triangular holes which occur between regions of the same fill region type are available for new placement by all three types. The holes within the red region, for

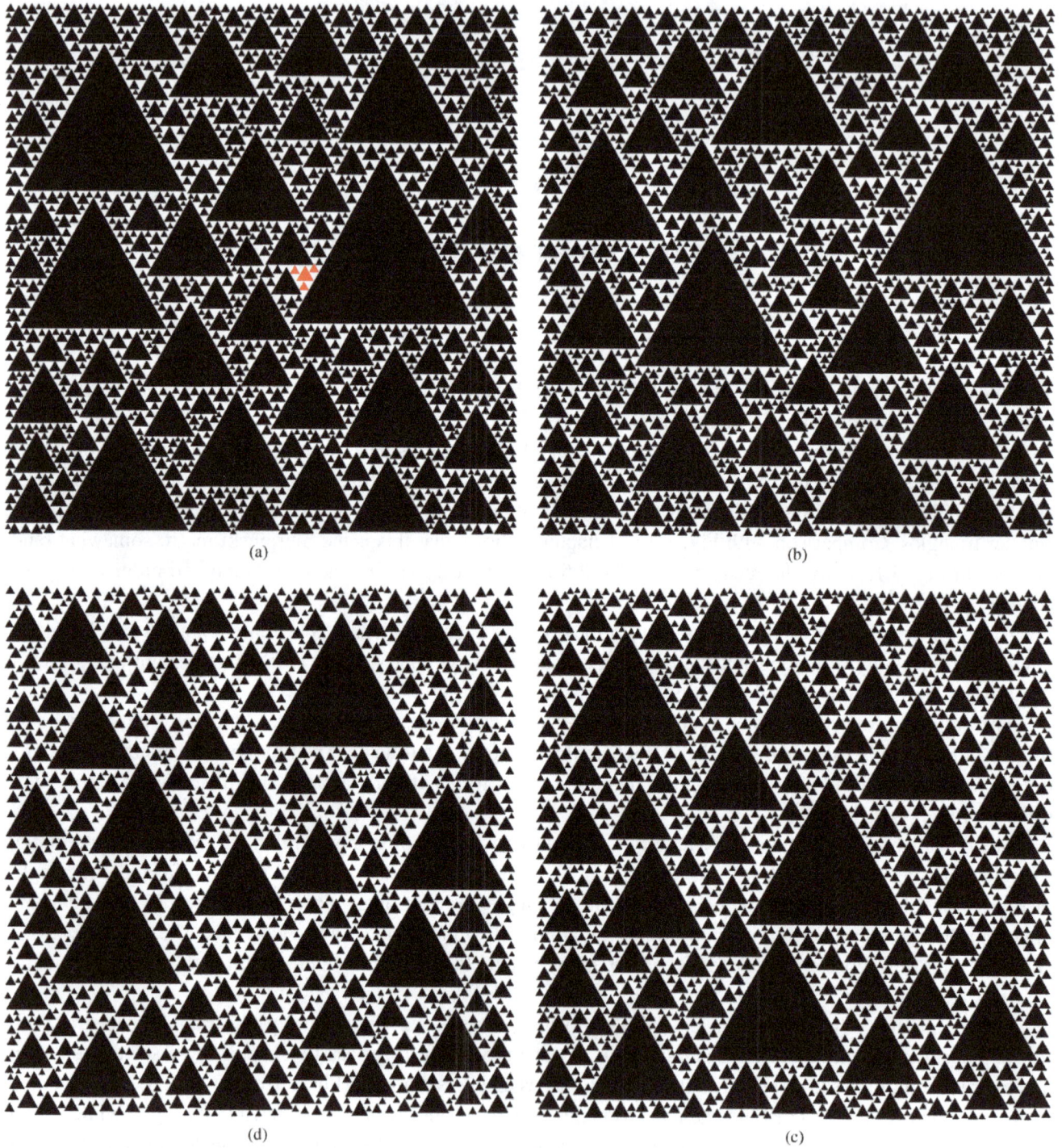

Fig. 3.1.   Randomly fractalized equilateral triangles. Figures (a)–(d) run clockwise from the upper left. In all cases there are 1200 triangles and $N = 4$. (a) $c = 1.24$, (b) $c = 1.21$, (c) $c = 1.18$, and (d) $c = 1.15$.

example, tend to favor more red diamonds there. The maximum c value is remarkably high here, nearly as high as one can attain with squares.

In Chapter 1, we saw that the corresponding tessellation, when suitably shaded, gives a strong three-dimensional illusion (Fig. 1.1(b)). It can be seen that the illusion persists here in the fractalization of the same

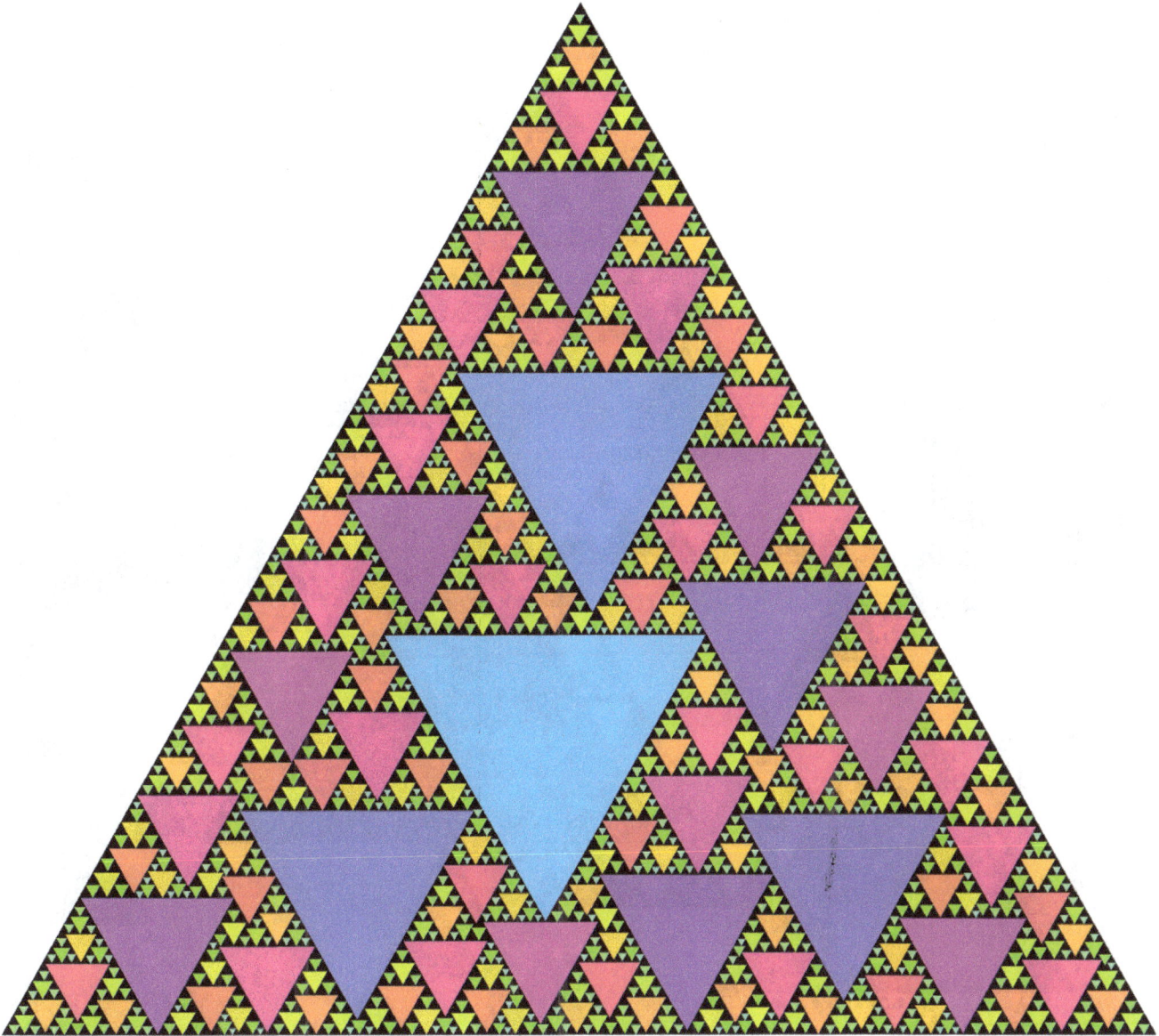

Fig. 3.2.   Triangles fractalized within a triangle. $c = 1.24$, $N = 3$, 1200 triangles, and 77% fill. Log-periodic color.

diamonds. The color scheme of Fig. 3.5 has been chosen to enhance this illusion and to many this image suggests an abstract set of buildings in some severe post-modernist style.

The examples in this chapter are among the simplest polygons. All except single triangles have high maximum c values and can achieve high fill factors. The reasons for this appear to be closely related to the shapes of the gasket holes. Straight, parallel sides are favorable for easy filling since they allow the placed fill region shapes to assume a number of positions along a vacant line. This is particularly so for squares, where one can see lines of squares of nearly the same size filled into the narrow gaps between large, early squares. Fill region shapes which create simple, easy tessellations (such as squares and rectangles) are also favorable for easy fractal filling.

Fig. 3.3. Equilateral triangles with two opposite orientations. The fill region shape alternates at each placement so that there are equal numbers of both types. $c = 1.40$, $N = 4$, 5000 triangles, and 94.5% fill. Red gasket. Because the white triangles at the top of the image blend into the white page background, the "dents" along the top edge are to be thought of as white triangles.

Fig. 3.4. Triangles with two 45° and one 90° corner. The four fill region shapes alternate cyclically at placement. $c = 1.39$, $N = 4$, 3300 triangles, and 93% fill. Each fill region shape has a different color. White gasket.

Fig. 3.5.   Diamonds with 60° and 120° corners. The three fill region shapes alternate cyclically at placement. $c = 1.54$, $N = 8,750$ diamonds, and 92% fill. The diamonds are colored red, light beige, and medium beige. The gasket is a dark beige.

Chapter 4

Discs and Circular Arcs

With discs, we begin to look at fill regions with curved boundaries. The maximum c values for discs are high, but do not go as high as the most favorable shapes having straight-line boundaries. The smallest discs in Fig. 4.1 typically have three near neighbors, which is much the same as for the Apollonian circles of Fig. 1.4 in Chapter 1. Another common *motif* is for the adjacent "next size up" neighbors of the small discs to be bigger discs where these bigger circles have four small near neighbors having radii about 1/4 of the larger one; this pattern is brought out by the log-periodic color scheme where these four neighbors have about the same color.

We do not show an example of multiple c values for discs (as in Fig. 2.7 of Chapter 2 or Fig. 3.1 of Chapter 3) but the basic rule of order at high c and disorder at low c is valid here too.

Figure 4.2 shows what happens when the algorithm is run repeatedly. Not surprisingly the discs are in different positions. What is interesting is that people's visual perception of these patterns tends to be "all the same". The fine differences in detail from one example to another make little impression on the visual sense. One can imagine these as sets of tiles given to eight different craftsmen, with instructions to place them in a circle "anywhere you want" starting with the largest and working down. The craftsmen might, of course, do things that are quite improbable but possible, like gathering all of the greens in one place, all of the blues in another, etc.

The annular ring fractalization shown in Fig. 4.3 was one of the "aha!" moments in the development of our subject. The original object was to seek shapes that cannot be fractalized and it was hoped that this would fail, revealing the limits of the algorithm. Instead, the algorithm ran smoothly and gave this. It has very strong correlation in the form of *nesting*. If you take the sequence of radii of concentric rings within one of the larger rings, you find an approximate geometric progression. The largest ring, for example, has nine quasi-concentric rings.

Another way to see the self-similarity is to look at all of the red rings with outer diameters about 1/8 of the image width. If you look at the rings within them, you will see "on average" nearly the same sequence of colors for the smaller rings indicating that they have about the same sizes in the same sequence. We do not see identical size sequences because of the competition between randomness and order.

The ratio of outer to inner diameter was chosen to be 2/3 in the example shown, but this ratio can be varied over a wide range. If it is quite small (say, 0.5 or below), the algorithm runs fast and can operate to a high c value. If the ratio is close to 1, the algorithm has to work very hard to make the next placement, so that c values are low and run times long. The rather sparse rings that one gets with ratios close to 1 will also require an N value substantially higher than 1 in order to allow the first few rings to fit into the area to be filled. Unlike the previous examples, this "donut" geometry results in a gasket which is separated into a huge number of pieces.

Two circular arcs can be combined to make a lens shape as shown in Fig. 4.4. This fill region fractalizes easily. This rather fanciful and surrealistic way of drawing the fill regions illustrates some of the possibilities for using fractalized shapes as art.

Fig. 4.1.   Discs fractalized in a square. Log-periodic color, white gasket. $c = 1.48$, $N = 3$, 2500 discs, and 96% fill. The average number of trials per placement was about 120,000.

In Fig. 4.5, an elongated lens shape has been fractalized with a random choice of the orientation angle at each trial. This results in strong angular correlation in the form of clusters with similar orientation (i.e., similar color). The same effect can be seen in Fig. 2.6 of Chapter 2. The large first-placed fill regions strongly constrain the orientations that later-placed fill regions can have. We see that the available holes in the gasket favor this. This image has periodic boundaries, so when a fill region is cut off at the left, the rest of it appears at the right, etc.

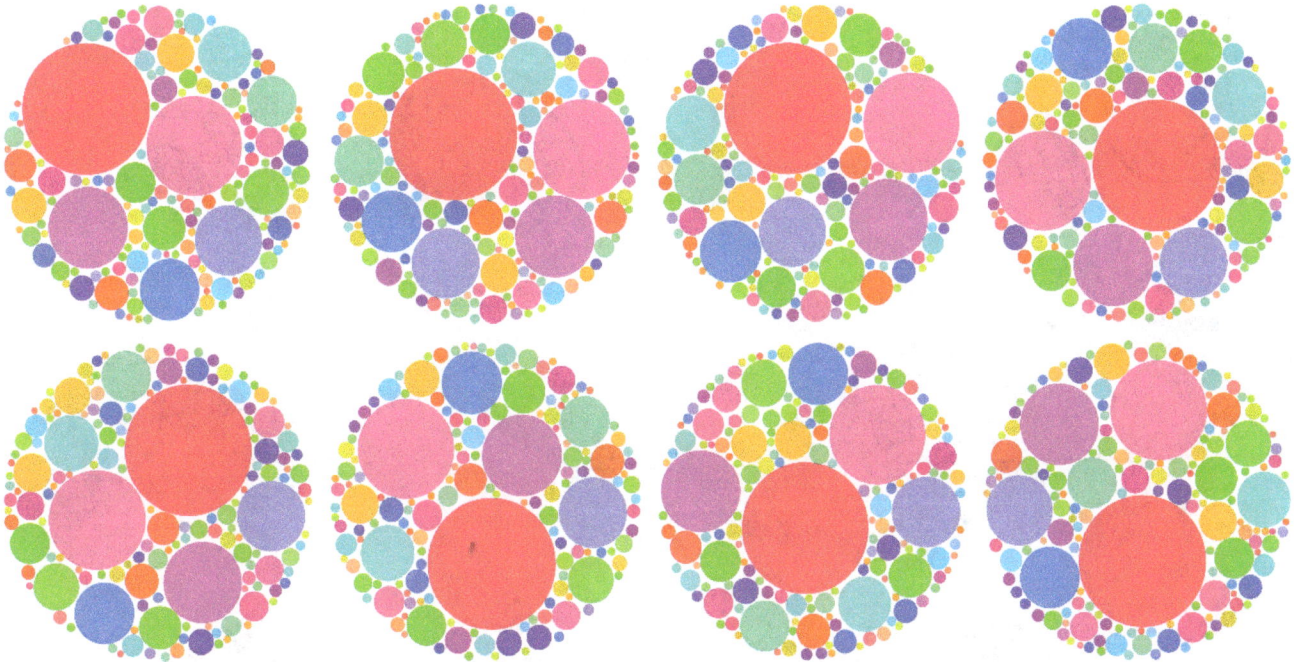

Fig. 4.2.   Repeated fractalizations of the same set of discs, using the same color sequence.

Two circular arcs can also be used to make the crescent shape shown in Fig. 4.6. In spite of the sharp cusps, this shape fractalizes easily, but does not allow high c values. There is strong correlation in the form of nesting, as also seen with the annular rings.

We have two circles paired side-by-side in Fig. 4.7. The eye perceives strong ordering, although it is difficult to precisely describe it. An interesting feature is the in-pointing cusps where the two circles meet, where we see a quite regular pattern of smaller bidiscs.

If Fig. 4.8 is rotated 90°, ∞ becomes 8. Close study will show correlation in the form of nesting, as seen with the annular ring fractal of Fig. 4.3. Some viewers report seeing faces. This is a rather sparse shape and the c value used here is about as high as it can be pushed.

The quad-disc is a quite sparse shape and the c value of 1.15 in Fig. 4.9 is about as high as it can go. It shows some impressive correlation despite the low fill factor. If you look at the largest quad (magenta color), you will see nested inside it, in roughly centered positions, smaller quads in purple, green, purple, and magenta. Many other examples of this can be seen.

Fig. 4.3.   Annular rings fractalized. Log-periodic color, black gasket. The inner diameter is 2/3 of the outer diameter. $c = 1.25$, $N = 2$, 2500 rings, and 84% fill. The average number of trials per placement was 16,000.

Fig. 4.4. Fractalized two-arc fill regions drawn as eyes. Black gasket, inclusive boundaries. $c = 1.288$, $N = 5$, 1500 eyes, and 81% fill.

Fig. 4.5.   Fractalized two-arc lens shapes with variable orientation. The orientation angle is randomly chosen at each trial. Periodic boundaries, black gasket. $c = 1.33$, $N = 2$, 1500 fill regions, and 90% fill. The color is a periodic function of the orientation angle, repeating every 180°.

Fig. 4.6.   Fractalized crescents. Log-periodic color, black gasket. $c = 1.24$, $N = 5$, 1200 crescents, and 74% fill. Inclusive boundaries.

Fig. 4.7.   Fractalized bidiscs; a two-disc fill region. Log-periodic color, black gasket, inclusive boundaries. $c = 1.39$, $N = 3$, 1200 bidiscs, and 91% fill. The average number of trials per placement was 80,000.

Fig. 4.8.　A four-circle fractalization of the infinity symbol. $c = 1.18$, $N = 2$, 2400 fill regions, and 73% fill. The average number of trials per placement was 13,000.

Fig. 4.9.   Quad-discs. The four discs are constrained to the square arrangement shown, with the distance between discs equal to the diameter of the discs. $c = 1.15$, $N = 2$, 1000 quad-discs, and 62% fill. The average number of trials per placement was 6000.

Chapter 5

Mixed and Irregular Shapes

It is possible to mix several fill region shapes. It is not necessary that the shapes all be congruent or similar to each other. One can fractalize a sequence in which all of the shapes are different provided that they follow the area rule (Eq. (7.1) in Chapter 7).

In this chapter, we go beyond the shapes of elementary geometry. In order to fully define the blob shapes of Fig. 5.5, we give an equation. The casual reader can simply accept the shape and skip the equation and its explanation with little loss. There is enough detail that those wanting to duplicate the work can do so.

Some polygonal mixed-shape fractals have appeared in previous chapters: Fig. 2.8 (2 shapes) in Chapter 2, Fig. 3.3 (2 shapes), Fig. 3.4 (4 shapes), and Fig. 3.5 (3 shapes) in Chapter 3. Here we consider more examples.

Figure 5.1 shows mixed discs and squares in which the choice of disc or square alternates at each placement. Trials were made with a disc until placement was achieved; then trials with a square until placement was achieved, then a disc, and so on. This forces the numbers of discs and squares to be the same. The sequence of areas was the same as it would be for either fill shape alone, with the disc radius r or square side half-length s calculated from the area by $r = \sqrt{A/\pi}$ or $s = \sqrt{A/4}$, respectively.

In Fig. 5.1(a), we see an example with relatively high c and find that there is strong clustering correlation. The discs are mostly adjacent to other discs and vice versa. The squares are easier to place than the discs; a total of 649,000 trials were made for disc placements, while only 542,000 were made for square placements, even though the numbers of the two shapes were the same. The ratio of these numbers comes closer to 1 as c decreases. In Figs. 5.1(a)–5.1(d), there is a progressive drop in the power-law exponent c and correspondingly weaker correlation. There is clustering of the two fill region shapes in all cases, but it diminishes as c falls. Order moves toward disorder as c falls toward 1.

It can be seen that the clustering order is not complete even for $c = 1.40$, and that each fill shape forms small islands within the other-shape areas, with these islands occurring on all length scales.

Figure 5.2 shows a mixed disc–triangle fractalization. As in the disc–square case, the choice of fill shape alternates at each placement. The statistics for this run showed that triangles are much harder to place and require about 100 times more trials than discs. From a purely visual point of view, it is hard to see strong correlation here. The parameter value $c = 1.31$ is higher than one can achieve with triangles only, but lower than the maximum c for discs only.

Figure 5.3 shows a mixed square–diamond fractalization. The shapes alternate sequentially at placement, ensuring equal numbers of squares and diamonds. The two shapes are related very simply here: the diamond is just a square rotated by 45°. There are several remarkable features. The c value is quite high and it appears that we can attain the same high c values as for squares only. We have seen examples of clustering correlation in other cases, but here it is quite extreme and can be described qualitatively by "squares and diamonds don't mix". Because of this ultra-clustering, we see a distinct "coastline" between the white and black regions with capes and bays at all

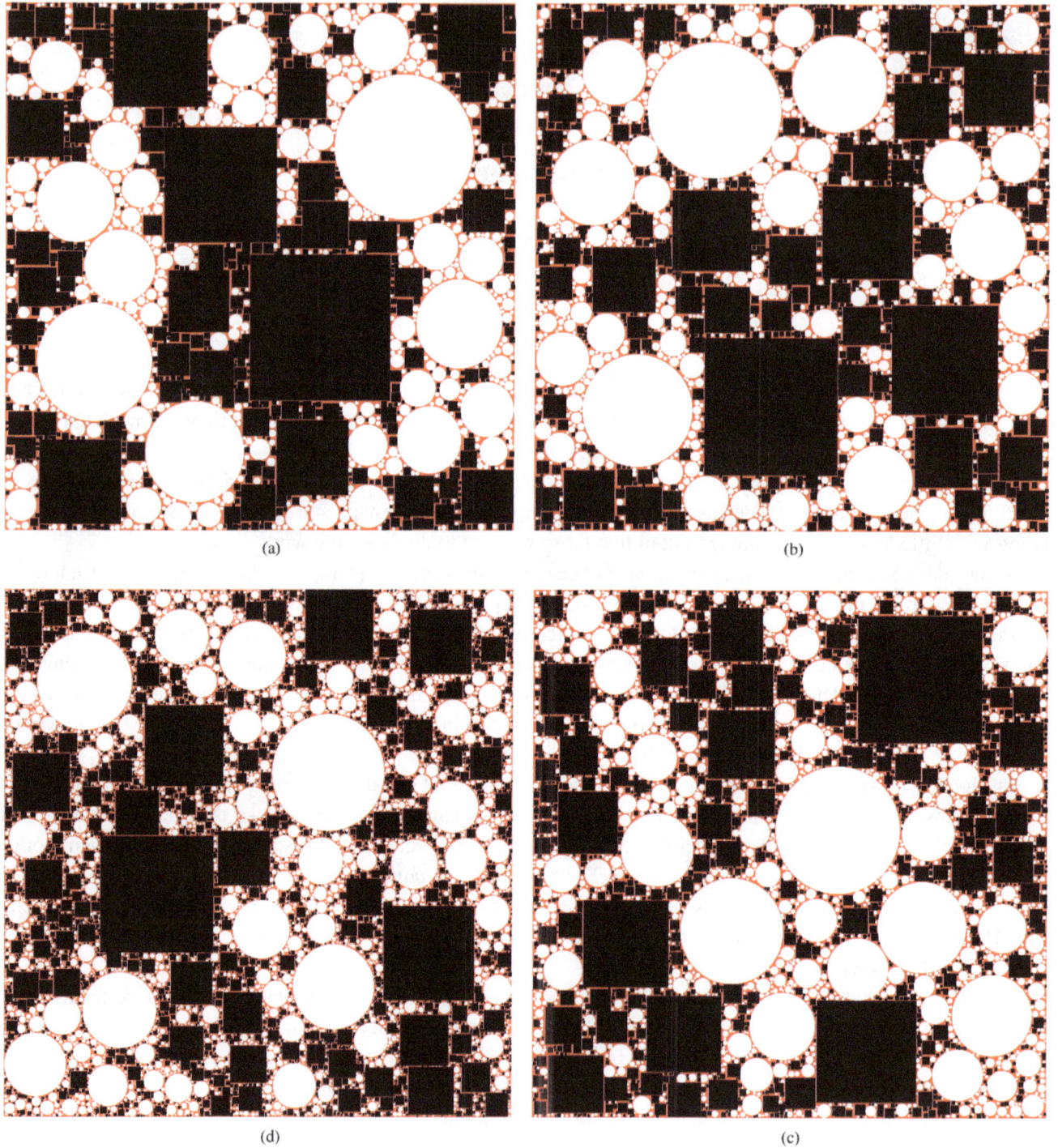

Fig. 5.1.   Mixed disc–square fractalization with the two fill shapes alternating at placement. (a)–(d) are clockwise from upper left. Equal numbers of the two fill shapes, red gasket. (a) $c = 1.40$, $N = 5$, 1442 fill regions, and 90% fill. (b) $c = 1.35$, $N = 5$, 1927 fill shapes, and 88% fill. (c) $c = 1.30$, $N = 5$, 3172 fill shapes, and 86% fill. (d) $c = 1.25$, $N = 5$, 7001 fill shapes, and 84% fill.

length scales. The coastline will have a fractal dimension of its own. One reaction to this image is that "It looks like one of those black-and-white dairy cows."

The observed behavior of the squares–diamonds fractal can be understood in terms of the gasket holes. With the relatively large N value used, the first few shapes placed have lots of room and much randomness. These first

Fig. 5.2.   Mixed disc–triangle fractalization with the fill shapes alternating at placement. Equal numbers of the two fill shapes. Red gasket, inclusive boundaries. $c = 1.31$, $N = 3$, 1000 fill shapes, and 84% fill. There were 7,954,000 triangle trials and 77,000 disc trials; a ratio of 103:1.

few shapes establish the "territory" claimed by squares and diamonds. The interstices between diamonds will have 45° angled sides and thus favor the placement of more diamonds there, and vice versa for squares. The only "unclaimed territory" is at the boundary between black and white where triangular holes offer equal opportunity for square or diamond placement.

Fig. 5.3.   A mixed square–diamond fractalization with shapes alternating at placement so that there are equal numbers of the two shapes. Gasket 50% gray. $c = 1.50$, $N = 20$, 4000 shapes, and 93% fill.

Patterns similar to squares and diamonds are seen in three dimensions in a mixed fractalization of hexahedrons and octahedrons. When c is near the upper limit, new octahedrons are placed preferentially near other octahedrons and conversely for hexahedrons, resulting in distinct regions of mostly one fill region shape.

Figure 5.4 shows how the square–diamond correlation effects vary with c. We see the usual diminishing order as c falls. In Fig. 5.4(a), there is near-total separation of the shapes, while as c falls, the squares and diamonds begin to intrude into each other's territory. When $c = 1.20$, there is a large amount of mixing.

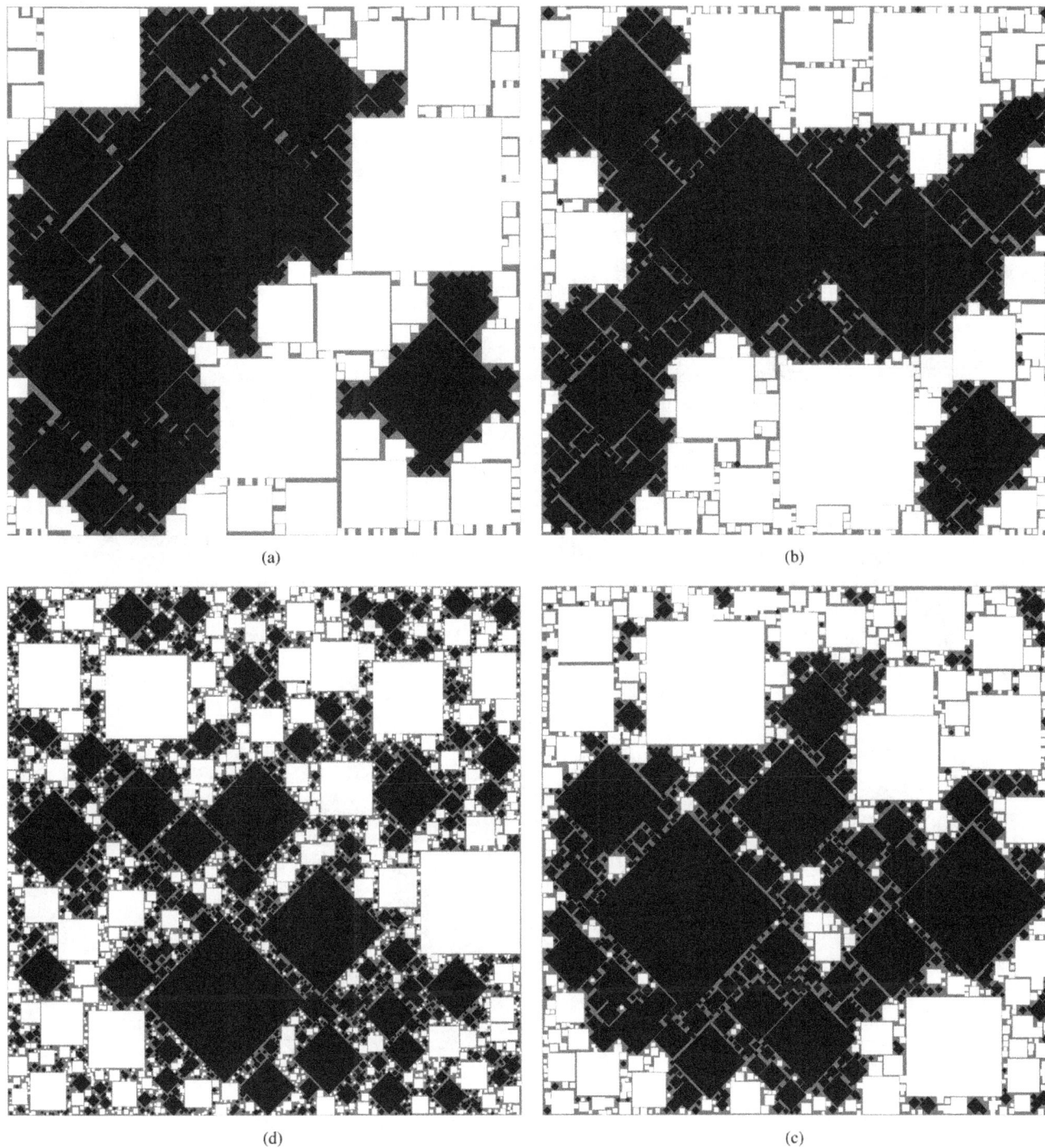

Fig. 5.4. Mixed square–diamond fractalizations with equal numbers of the two shapes and different c values. Gasket 50% gray, inclusive boundaries. Figures (a)–(d) are clockwise from upper left. (a) $c = 1.50$, $N = 4$, 348 shapes, and 90% fill. (b) $c = 1.40$, $N = 4$, 701 shapes, and 88% fill. (c) $c = 1.30$, $N = 4$, 1288 shapes, and 83% fill. (d) $c = 1.20$, $N = 4$, 6815 shapes, and 78% fill.

The blobs of Fig. 5.5 are defined by Eq. (5.1) in polar coordinates. These blob shapes are *all different*, yet they fractalize readily with the usual sequence of areas. The casual reader can skip the following equation and just enjoy the blobs.

$$r(\theta) = R\{1 + \delta \left[\cos(2\theta + \phi_2) + \cos(3\theta + \phi_3) + \cos(4\theta + \phi_4)\right]\} \tag{5.1}$$

Fig. 5.5.   Fractalized blobs. The fill region shape is randomly varied, and no two of them are congruent or similar. Blobs with the same color have the same shape. Black gasket, periodic boundaries. $c = 1.32$, $N = 1,400$ fill regions, and 88% fill.

The relative smoothness of the edge can be varied by changing δ. When $\delta = 0$, we have $r(\theta) = R$ which is the equation of a circle. As δ is increased, the edge gets more "blobby", the maximum c value drops, and the average number of trials needed to place a shape increases sharply. It is concluded from this that congruence or similarity of shapes is not a requirement of the algorithm.

Many of the fractalized shapes here are common symbols. There is something to be gained by using a shape that is recognized by the eye as having some meaning. An attractive idea is to fractalize mixed + and − symbols as in Fig. 5.6. The practical problem is that it is quite hard to design two such symbols which look like typical

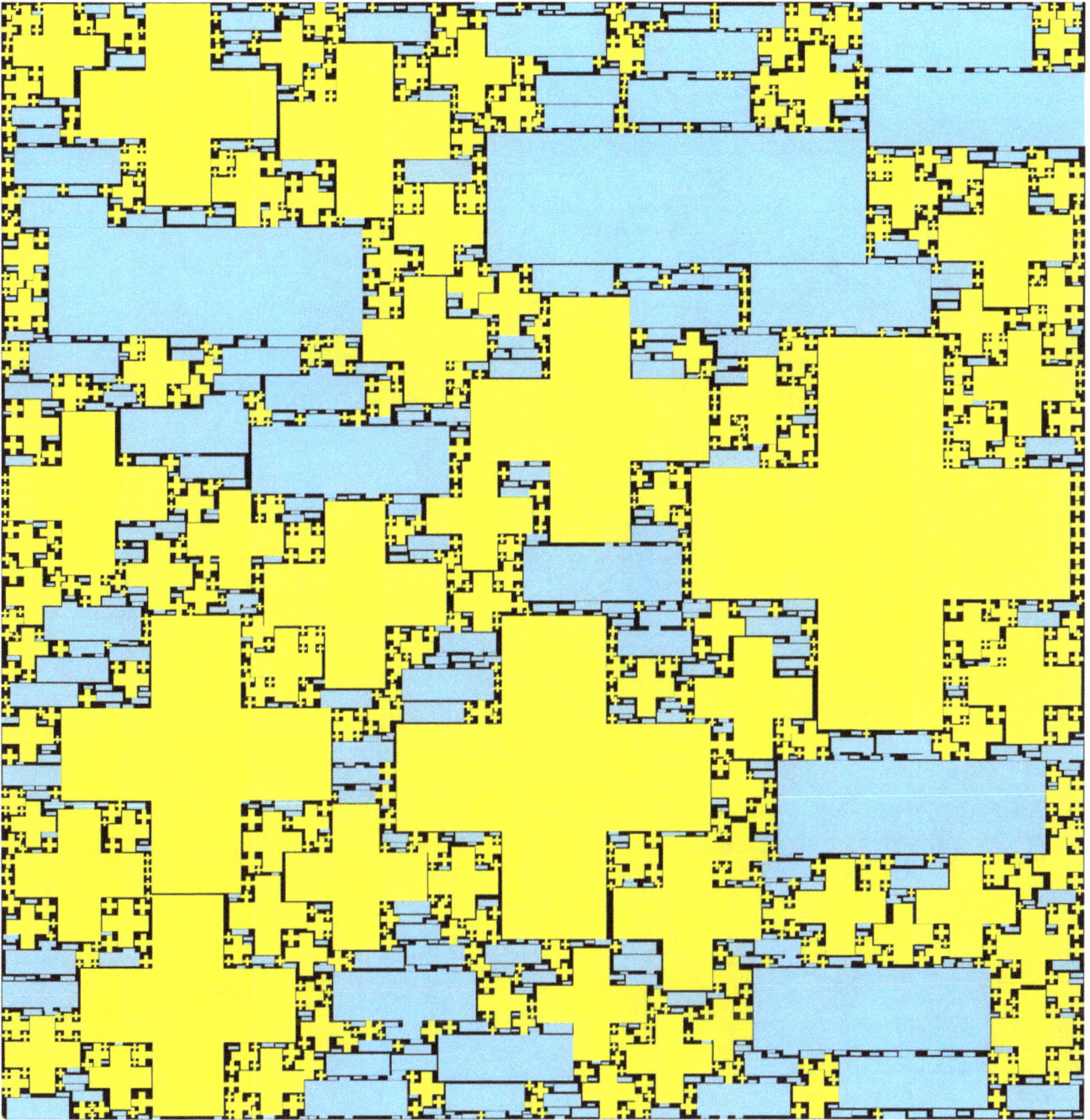

Fig. 5.6.   Fractalized plus and minus symbols. The shapes alternate at placement. The areas of the two symbols are in a 5:3 ratio. Black gasket, inclusive boundaries. $c = 1.35$, $N = 3$, 2100 shapes, and 88% fill. The average number of trials per placement was 21,000.

examples and still have the same area. One way around this is to have the minus symbols be smaller at a given stage in the algorithm. How would one do this? In Fig. 5.6, we have used *paired* symbols. An area sequence of the usual kind is generated. At each step, we place two shapes, a plus with 5/8 of the given area and a minus with 3/8 of the area. This ensures equal numbers of the two symbols. This paired placement process remains space-filling, but gives additional flexibility in the two shapes.

Figure 5.6 shows the fractalized plus–minus pattern. There is quite strong clustering, but in considering the visual effect, it needs to be kept in mind that if the algorithm were run "to infinity", 5/8 of the area would be yellow, and 3/8 blue.

Chapter 6

Examples

The view taken here is that "an example is worth a thousand words". One of the claims made is that one can fractalize with a fill region of any shape. While a formal proof of that seems remote, a large number of examples makes it plausible. The examples given here have both mathematical and artistic features. The application of the algorithm to art has been treated in several papers by Dunham and Shier [9–11].

Figure 6.1 shows fractalization of two semicircles on the left. One has the flat side up, the other the flat side down. The artist can ask "What recognizable objects can these be?" The answer illustrated here is that they can be eyes and mouths. One can assign a surrealistic title like "How Dali remembered the girlfriends of his youth".

The targets of Fig. 6.2 are constructed using three circles, and c is close to its upper limit. The radii are chosen such that six smaller targets can fit exactly in the annular space within a target. What is interesting is that in most

Fig. 6.1. Fractalization of semicircles in both geometric and artistic aspects.

Fig. 6.2.   Fractalized targets. Log-periodic color. The inner radii are 1/4 and 3/4 of the outer radius. White gasket, inclusive boundaries. $c = 1.19$, $N = 3$, 3000 shapes, and 74% fill.

cases there are six large targets in this space, with radii only slightly smaller than the "just fits" radius. This is seen to be the case at all length scales, which illustrates the idea of "statistical self-similarity". The smallest targets are mostly either filled with next-size targets or empty. A corresponding fully-ordered recursive target fractal could be constructed where a single large target has six exactly-fitting smaller ones, etc. in the manner of the Apollonian circles of Fig. 1.4 in Chapter 1.

The shape of Fig. 6.3 is quite sparse. It consists of a circle with seven small discs with 1/3 of the outer radius cut out of it. The c value used here is about as high as it can be pushed. This shape has about the lowest maximum

Fig. 6.3.   Fractalized "seven-holes". Log-periodic color. White gasket, periodic boundaries. $c = 1.079$, $N = 5$, 1000 shapes, 34% fill. Each time the fill region shape is drawn, the holes are colored.

c value of any that have yet been studied. By a suitable choice of the period for the log-periodic color, one can make all of the inner features of a given seven-hole about the same color.

Figure 6.4 is another quite sparse shape. Snowflakes are popular on winter clothing. Readers familiar with crystallography will appreciate that all of the angles are multiples of 30°. With images of this kind one must limit the number of fill regions if they are to be resolved in the finished print. From the viewpoint of art, the wintry sky is just as important as the snowflakes in setting the scene.

Fig. 6.4.   Fractalized snowflakes. Blue gasket, inclusive boundaries. $c = 1.08$, $N = 3$, 500 shapes, and 34% fill.

Figure 6.5 illustrates a variation on the basic algorithm. The orientation angle of the squares is a smooth, continuous function of x and y. This is accomplished by using a sum of a small number of periodic sinusoidal "harmonics" for the angle variation. We desire to have a high fill, but don't want the biggest and smallest squares to be too different in size. High c and N does this. This image is an intimate blend of randomness and order. It could be an aerial view of some long-established city that grew up higgledy-piggledy from cow paths. Curving streets seem to flow between the houses. Since it has periodic boundaries, it could be tiled for use as a fabric design or a wallpaper design.

Fig. 6.5.   Flowing squares. $c = 1.57$, $N = 150$, and 1250 shapes. Black gasket, periodic boundaries.

The algorithm is able to do interesting things with text. Figure 2.10 in Chapter 2 shows an example where the text is the fill region. In Fig. 6.6, we have used text as the bounding region, with various fill region shapes. The fill region shape can be chosen to match or emphasize the meaning of the word, as in AMOR. Because of the random nature of the statistical geometry algorithm, the same letter is always somewhat different. This can be seen in the two letters I in VINCIT. One must use a high value of N in order to get the first fill region to fit into the narrow bounding region.

Fig. 6.6.   A font with fractal filling. This is a Latin text from the Roman poet Virgil, and means "Love Conquers All". It is sometimes cited as Omnia Vincit Amor meaning "All are Conquered by Love".

Like tessellations and Sierpinski's construction, random fractals can make images that are inherently engaging to the eye. The possible shapes include symbols, birds, and animals. The rest of this chapter explores the possibilities in both geometric and art aspects.

The large number of shapes which are shown provide evidence that the algorithm works for any shape or sequence of shapes. No exceptions have yet been found.

Many of these fill region shapes recapitulate ones found in the previous chapters. Example 4 shows the use of a simple rectangle fractal for patriotic purposes. The bug cars of Example 14 are popular with small boys. Example 13 shows how art can be applied to side-by-side ellipses to produce a sort of fractal family tree.

Here we move away from geometry and more toward art. Two kinds of hearts are shown (28 and 29). The sailboats (20) are actually irregular pentagons drawn as boats. The microbuses (19) will be a nostalgia item for some baby boomers. The butterflies (23) are defined using local polar coordinates. Any fill region shape that can be represented by a Fourier series $r(\theta)$ can be used this way.

It can be seen that color and contrast make a large difference in the viewer's perception. Example 32 is not one of the shapes of school geometry, but is popular. Examples 43 and 44 show how shading creates perceptions beyond the basic shape. Example 47 was rather bland until it was given the colors of hot peppers. Example 41 illustrates the confusing directions often encountered in contemporary life.

Most of the images have strongly contrasting boundaries (e.g., black and white). Example 49 shows what can be done with less contrast. The vehicles like Example 52 were popular with the author's preschool grandsons. His quiltmaker wife thinks the flower patterns should be fabric designs.

Letters and words in any alphabet are shapes, and fractalizing them is a test of the algorithm as well as a rather novel way of presenting text. Examples 54, 55, 62, 67, 69, and 70 show Latin letters, while 71, 72, and 74 show letters used in words. (Example 72 is "a thousand times NO!") Example 36 shows a Greek letter. Example 73 is the mystical syllable "om" in one of the alphabets of India. Example 75 is a mixed fractalization of the digits 0–9. Examples 7, 34, 48, and 76–79 show the use of hollow shapes. With the astroids of 65, the shape of the gasket is as interesting as that of the placed fill regions. The fill regions here are everywhere concave except at the cusps.

These are quite varied. Number 80 has the title "don't bug me". Number 82 was created for the author's guitar-playing eldest child. The "suns" image (84) can be an interesting test case for fractalization of sprawling shapes. Number 89 uses entirely circular arcs. Number 91 is a mixed-shape fractalization of the standard "operator" symbols of arithmetic: $+ - \times / =$. Number 95 is a sort of political cartoon. We see the red jaws munching in from the right and the blue jaws from the left. There is intra- and inter-color conflict at all size scales.

Numbers 96 and 97 show spears and arrows with random orientation and periodic boundaries; color is by angle. Number 98 begins a series (nos. 98–105) with ellipses, where they are used as "faces in a crowd". The ellipse is a particularly attractive shape because so many things have near-ellipse shapes. Number 107 shows groundhogs looking out of their burrows on Groundhog Day. Number 108 looks a lot like 11, with a similar color scheme, but it has four diamonds rather than three. The mushroom shape of 110 and 111 is easily recognized.

0

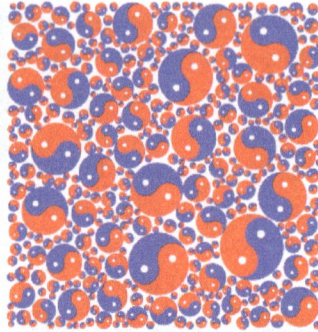
1

2

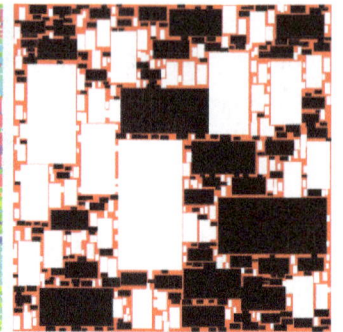
3

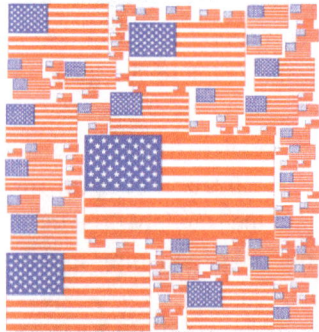
4

5

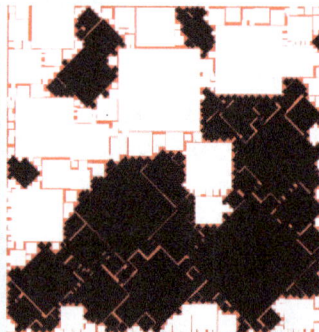
6

7

8

9

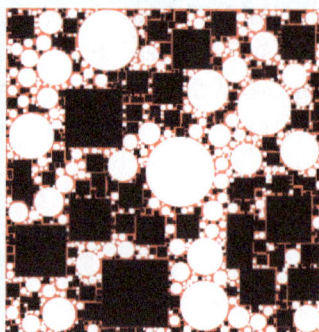
10

11

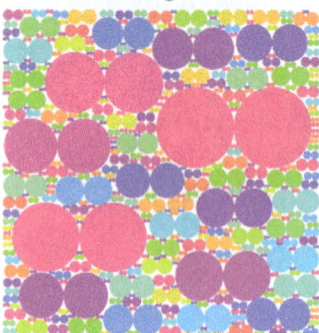
12

13

14

15

16

17

18

19

20

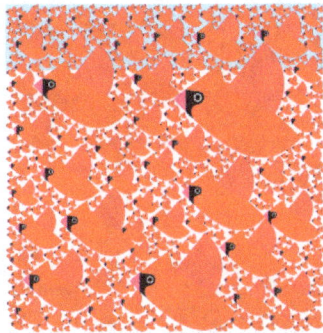

21

22

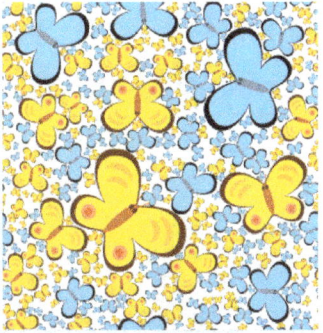

23

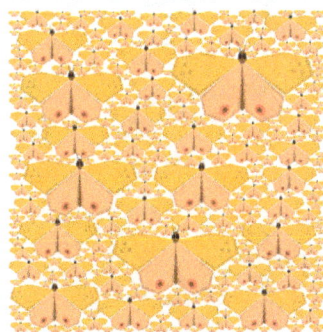

24

25

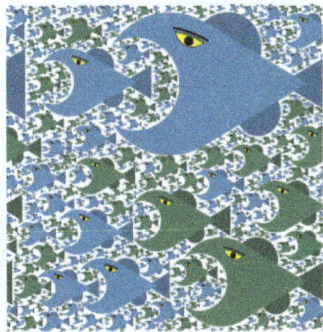

26

27

28

29

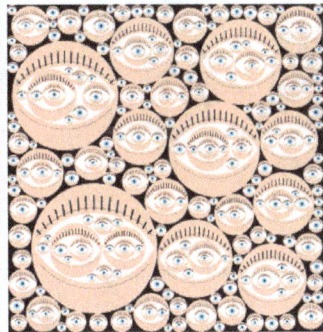

30

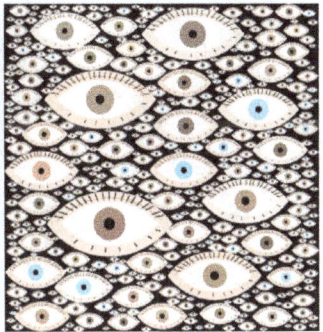

31

32

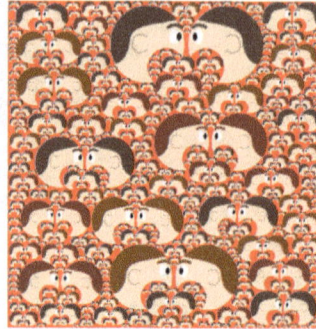

33

34

35

36

37

38

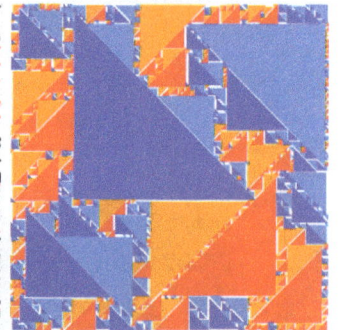

39

40

41

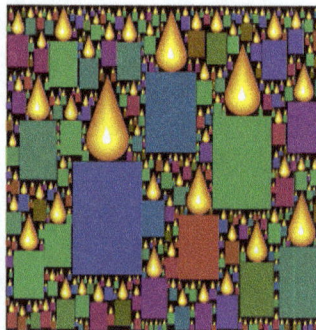

42

43

44

45

46

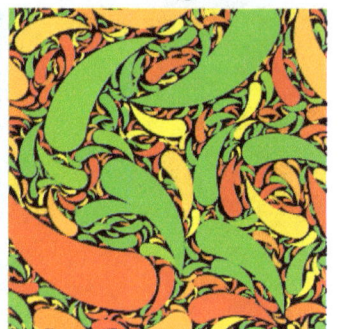

47

48

49

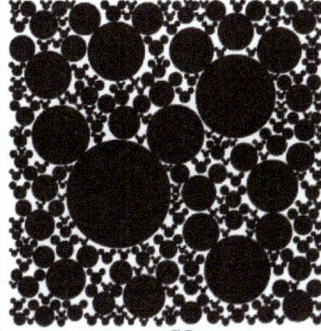

50

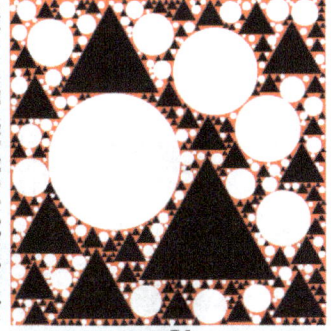

51

52

53

54

55

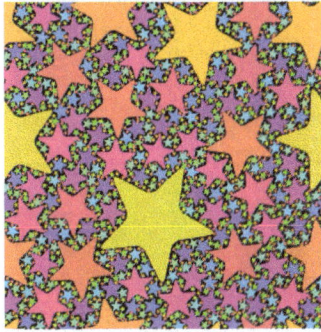

56

57

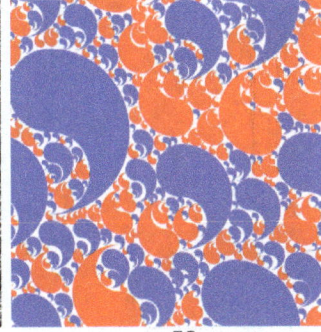

58

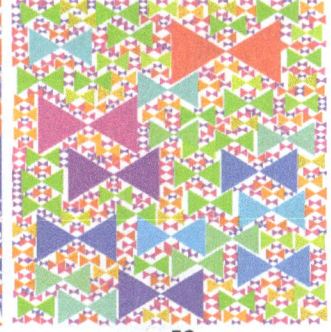

59

60

61

62

63

64

65

66

67

68

69

70

71

72

73

74

75

76

77

78

79

80

81

82

83

84

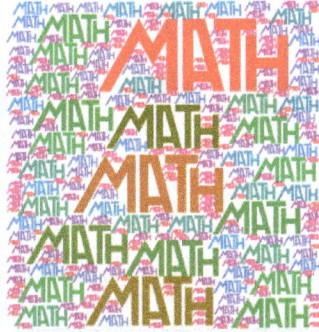

85

86

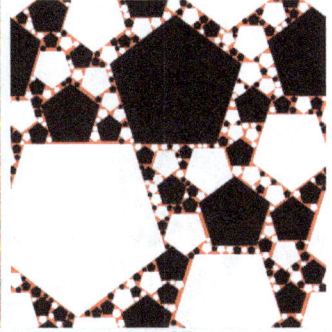

87

88

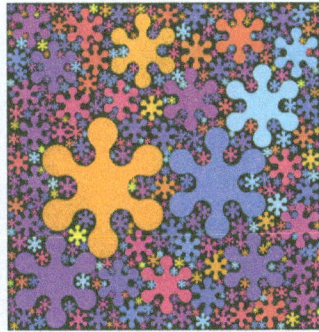

89

90

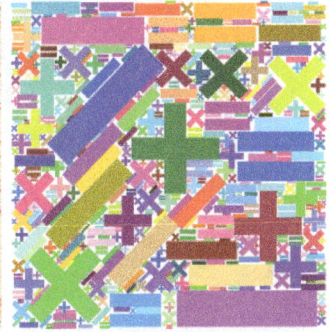

91

92

93

94

95

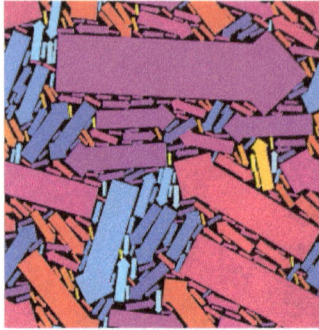

96

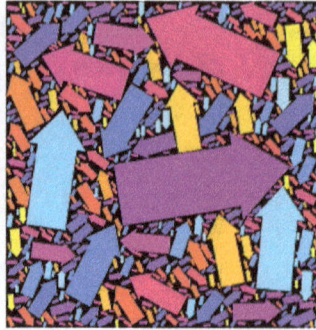

97

98

99

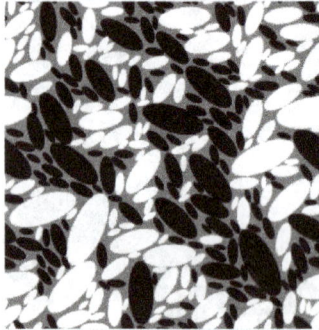

100

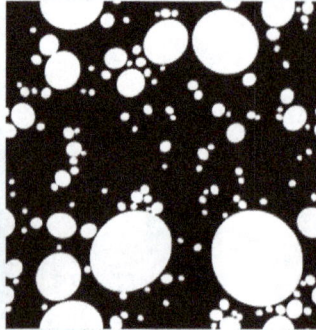

101

102

103

104

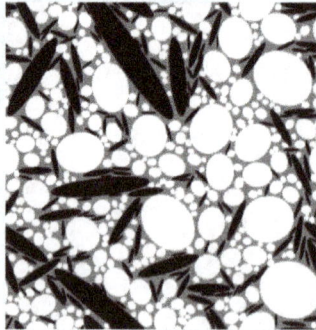

105

106

107

108

109

110

111

Chapter 7

The Statistical Geometry Algorithm

The focus in the first part of the book has been on describing images made by the statistical geometry algorithm, and formal mathematics has been avoided. We now turn to a mathematical description of the algorithm [7, 8]. We consider two Euclidean dimensions first because it has received the most attention and the images are easily shown on flat paper or screens. The algorithm works equally well in one and three Euclidean dimensions.

The basis for the conclusions in this chapter is *empirical data from computer runs*. Although the algorithm can be described in purely mathematical terms, rigorous proofs (or refutations) for most of the claims made here remain to be found. One such claim is that for any fill region shape there is a wide range of c values for which the algorithm never halts. From the viewpoint of pure mathematics this must be thought of as a conjecture.

7.1. The Algorithm Stated

Chapter 1 talked of "the standard sequence" of areas, and noted that it has two parameters c and N which define its properties. We denote the sequence of areas by $A_0, A_1, A_2, \dots$. These are the areas of the fill regions to be placed within the bounding region. Suppose that the bounding region to be filled has area A. The key result is the *area rule*:

$$A_i = \frac{A}{\zeta(c,N)(N+i)^c} \tag{7.1}$$

where $i = 0, 1, 2, \dots$. Here, $\zeta(c,N)$ is the Hurwitz zeta function defined by

$$\zeta(c,N) = \sum_{m=0}^{\infty} \frac{1}{(N+m)^c} \tag{7.2}$$

It follows from Eqs. (7.1) and (7.2) that

$$\sum_{i=0}^{\infty} A_i = \sum_{i=0}^{\infty} \frac{A}{\zeta(c,N)(N+i)^c} = \frac{1}{\sum_{m=0}^{\infty} \frac{1}{(N+m)^c}} \sum_{i=0}^{\infty} \frac{A}{(N+i)^c} = A \tag{7.3}$$

Thus, the sum of all the areas A_i equals the area to be filled, i.e., the algorithm is "space-filling in the limit". Convergence of the sum in Eq. (7.2) requires that $c > 1$ and $N > 0$. The parameter N is a positive integer in most of the examples in this book, but it can in principle be any real number greater than 0.

The properties of the Hurwitz zeta function are such that as $N \to 0$, the area for $i = 0$ completely dominates. Because of this there is a minimum value of N below which it is not possible to fit even fill region 0 into the bounding region.

Unless otherwise stated, the material in this chapter assumes two dimensions, i.e., the fractal pattern is planar.

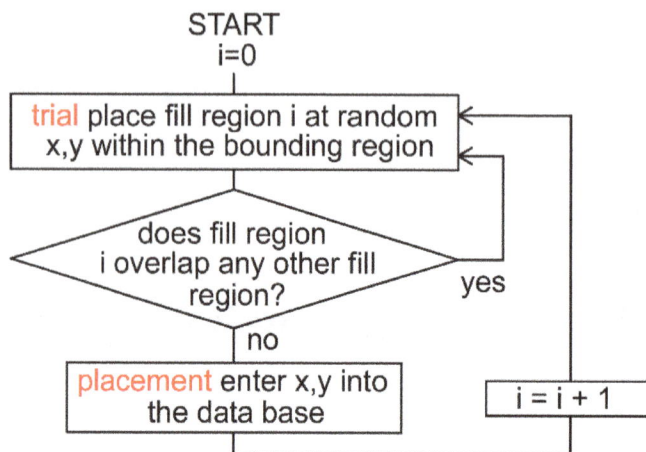

Fig. 7.1.   Flow chart for the statistical geometry algorithm. Trial and placement are defined.

The algorithm for making a fractal pattern of the fill regions can be illustrated with a flow chart. With the area sequence defined by the chosen values c and N, the algorithm proceeds according to Fig. 7.1.

Figure 7.1 represents a very simple flow chart. The step with $i = 0$ can be thought of as the *initial placement*, and is the only placement that does not require an overlap test with previously-placed shapes. The area rule does not involve randomness; chance only comes in during the trials and placements. The flow chart has no STOP point because all available evidence says it never halts. Practical computation requires a STOP.

The perceptive reader will see that we have referred to a fill region, but have nowhere said what its shape is. This is not an omission. Existing computational evidence indicates that *the algorithm works for fill regions with any shape or sequence of shapes* with areas obeying Eq. (7.1) (for some continuous range of c values) as seen in many examples in the previous chapters. This is a rather surprising claim, but despite much work no shape or shape sequence has yet been found that provides a counterexample. The blobs of Fig. 5.5 in Chapter 5 provide a particularly interesting case, as the shape of every fill region is different and none is congruent with any other.

The gasket region is a single connected whole in these constructions (unless the shape is hollow, which is treated in Sec. 7.10). If one assumes the use of random numbers having infinite precision, the probability of any two shapes touching is vanishingly small. For discs of radii r_1 and r_2, with centers at a distance r_{12}, the criterion for overlap can either be $r_{12} < r_1 + r_2$ or $r_{12} \leq r_1 + r_2$. The choice makes little difference in computation of images, but would require a clear definition in formal mathematical studies.

There is always some finite distance between the fill regions. In the *packings* of mathematics and physics [4–6], all of the fill regions are mutually tangent. For this reason, statistical geometry patterns are called *fillings* rather than packings. It is surprising that a pattern where none of the shapes touch each other can fill all the space in the limit, but available evidence says it is so.

Many examples of the algorithm are shown as images in previous chapters, with their values of c and N. The average number of trials per placement is often quite large, in the thousands or even millions.

Moving from one shape of the fill region to another, only two changes need to be made. One must find a formula for the linear dimension(s) of the shape as a function of its area, and one must create the code for a test which will determine if two shapes overlap each other. The development of code for testing overlaps is treated in Chapter 10.

The placed fill regions are numbered $i = 0, 1, 2, \ldots$ thus far. In some situations, it is desirable to number them by $n = 1, 2, 3, \ldots$. Thus $n = i + 1$. When log–log plots are used, the fact that $\log(0)$ does not exist is important.

7.2. Does the Algorithm Halt?

This is a central question. The algorithm is space-filling only if it does not halt. The observed behavior can be described in terms of a halting probability which varies with c and is shown schematically in Fig. 7.2.

For a substantial range of c values ($1 < c < c_1$), the process does not halt. For higher c values, the halting probability varies with c following an s-shaped curve as shown in Fig. 7.2, while if $c > c_2$, the algorithm always halts.

Much of the detailed numerical work described here assumes $N = 1$. Experience shows no fundamental differences when $N > 1$.

Figure 7.3 shows observed halting. Many runs are made, and the fraction of them which halt is shown in the figure. There are enough runs that this is reasonably accurate and it is seen that the data agree with the general scheme of Fig. 7.2. This kind of detailed study has not been done for most of the images shown in the book, and the "maximum c" referred to in earlier chapters is simply the highest c value at which runs rarely halted.

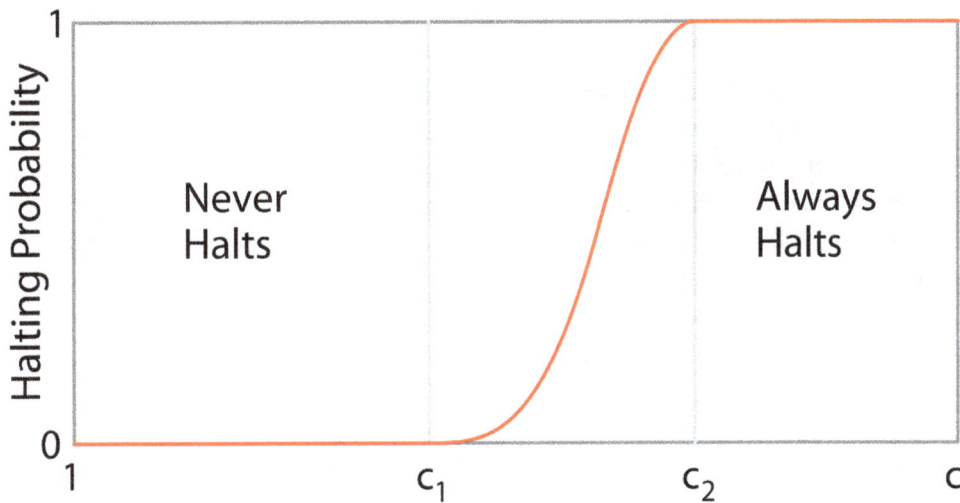

Fig. 7.2.   General scheme for the halting probability and its variation with c.

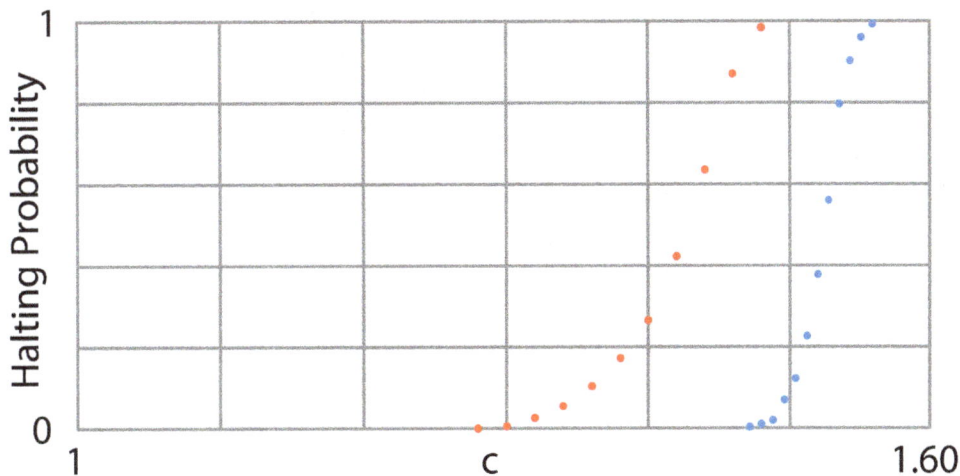

Fig. 7.3.   Computed halting probability data for discs in a square when $N = 1$. The red points are for inclusive boundaries (2000 runs per point). The blue points are for periodic boundaries (300 runs per point).

Statistical detail on the halting behavior and the number of placements which occur before halting can be found in Ref. [8]. Halting has an "infant mortality" trend. If a run is going to halt, there is a very great likelihood that this will occur during the early placements. The halting behavior is different for inclusive boundaries than for periodic boundaries.

The explanations of c_1, c_2 and "infant mortality" in two dimensions are mostly qualitative at the present time. Squares provide a simple example. If we consider the placement of squares within a square, the first two squares get larger and larger as c increases. If s_0 and s_1 are the side lengths of the first two squares and s is the side length of the bounding square, it becomes impossible to place square 1 if $s_0 + s_1 > s$. Numerical calculations show that this occurs when $c > 1.5224$ and $N = 1$, thus setting an upper limit for c. The computationally observed behavior of square fractalization runs agrees with this. It will be seen that as c approaches 1.5224, the random choice of placement for the first two squares must be the highly improbable one of the two squares being very near to diagonally opposite corners. If a high-c run fails to meet this, it will halt at the placement of square 1.

A worst-case analysis shows that for $N = 1$ for discs in a circle, there are always random positions of discs 0 and 1 which can lead to halting if $c > 1.20977$. For this case, the c value above which the first two discs cannot both fit is 1.5224 (the same as for square-in-square). This suggests that $c_1 = 1.20977$ and $c_2 = 1.5224$ for discs in a circle with $N = 1$.

The worst-case inclusive placement of early large squares in a square is easy to see for $N = 1$. When $N > 1$, the first few shapes are closer to the same size and determination of the worst-case arrangement is much more complicated.

Discs in a square present a somewhat more complicated situation for the placement of the first few discs. It is actually at the placement of the third disc that worst-case fit occurs, as seen in Fig. 7.4. The range of x, y trials for the first three discs allowing such a configuration will be an exceedingly small fraction of all possible trials, indicating that the probability of such an event is quite low. In practice, it is found that squares can be placed up to higher c values than discs (within squares in both cases) before halting becomes a serious problem. When discs are placed in a circle ($N = 1$), the first two discs cannot both fit when $c > 1.5224$, showing that shape of the boundary plays a role here.

Constructions such as Fig. 7.4 are relatively easy to make for inclusive boundaries. The corresponding constructions for periodic boundaries are much more involved since (for example) a disc which is placed at one of the corners has three more periodic "partners". Figure 7.3 shows that there is much more freedom and "slack" for the early placements when periodic boundaries are used.

Fig. 7.4.   Worst-case high-c placement of the first three discs in a square when $N = 1$. The touching of discs 0 and 2 occurs when $c = 1.577$ (shown).

The observed "infant mortality" of halting [8] suggests that if a computer run can get past the first few placements, it will indeed go on indefinitely. For squares with inclusive boundaries, successful runs can be made for c values beyond $c = 1.50$ if one is willing to make many runs and wait for a successful one. (Alternatively, one can pre-place the first two fill regions at optimal locations.) For discs fractalized within a square boundary, the practical upper limit of c for successful runs is somewhat lower than for squares with such a boundary.

What effects account for c_1? Why does the algorithm continue without halting for "low" c values, as it is observed to do? This question is taken up in Sec. 7.4.

Summing up, it can be seen that a geometric analysis of the first few placements can account for the existence of c_2 as shown in Fig. 7.2. Because the algorithm is described entirely in mathematical terms, it is possible in principle to find an analytical means of determining the exact shape of the curve of halting probability versus c between c_1 and c_2.

For different fill region shapes c_1, and c_2, the s-shaped curve shift up and down; indeed *differences in the upper limit of usable c values are the main difference between one fill region shape and another*. Compact fill regions like squares and discs have the highest usable c values. Figure 4.10 in Chapter 4 shows a quad-circle fractal; this sparse and sprawling shape has a quite low maximum c value.

This process is a rather odd form of chance. In the form of coin tossing and many others, random chance usually gets one into trouble, and the longer one does it the greater the difficulties. Here we have a chance process which *always succeeds* however long one continues. Imagine a vendor of decorative tiles who creates a set of 100 colorful round tiles having the dimensions of the first 100 discs in a fractal which is set up for a bounding area of 1 m^2. This is offered to tile setters with the instructions: "Mark off an area of one square meter, either a circle or a square. Start with the largest tile, and attach it permanently anywhere you wish in the marked-off area. Continue to permanently attach the tiles anywhere you wish, proceeding always from larger to smaller. *There will always be a place for every tile regardless of how you choose to place them*." How many experienced tile setters would believe this? The pattern thus created would not be artless; the tile setter has an immense number of possibilities in the choice of colors and arrangements.

7.3. Run-Time Behavior

The conclusion from Eq. (7.3) is that the process is space filling *if it does not halt*. Detailed statistical studies (Fig. 7.3) strongly support the idea that there is a wide range of c where halting does not occur. How does a computer run of the algorithm behave under these conditions? Figure 7.5 shows the cumulative number of trials n_t needed for n placements in log–log coordinates.

Previous familiarity with log–log plots is of great help here. For those unfamiliar with them, we offer a brief explanation. Suppose we have a power–law relationship $x = ay^s$. If we take logarithms of both sides, this becomes $[\log(x)] = \log(a) + s[\log(y)]$, i.e., a straight line for $[\log(x)]$ versus $[\log(y)]$. The exponent s is the slope of the line.

The data in Fig. 7.5 gives a reasonable fit to a straight line; thus the number of trials can be well approximated by a power law versus the number of placements. That indicates that within the range of c values shown ($1 < c \le 1.40$ for discs and $1 < c \le 1.45$ for squares) the average number of trials needed to place n fill regions increases steeply with n but *is always finite*. The c values for the data of Fig. 7.5 lie mostly within the "no halt" region, but the upper data sets have c values above c_1. Within this c range, halting was rarely seen for discs or squares, which illustrates the fact that if "infant mortality" is not encountered, most runs continue indefinitely. There were no halted runs in the computations for Fig. 7.5 although there is a small probability that one would happen with the higher c values.

It has been a persistent observation that the log slopes of cumulative trials versus placement trends are quite close to $+c$. Further insight into this is offered in Sec. 7.4.

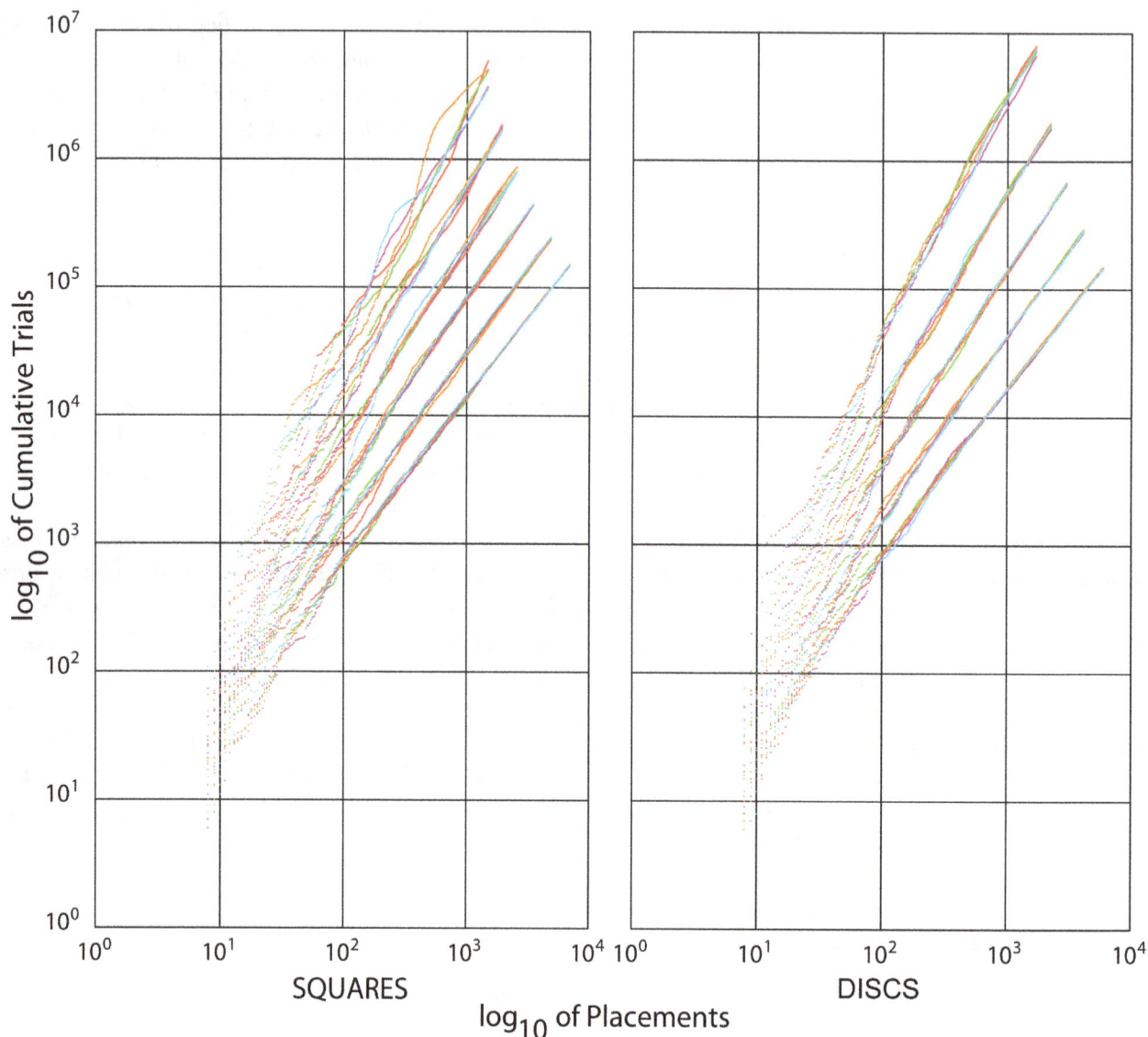

Fig. 7.5.   A log–log plot of the cumulative number n_t of trials needed to place n discs or squares in a square. $N = 1$ in all cases. There is data for five runs for each value of c. For the discs, the c values are (from the bottom) 1.20, 1.25, 1.30, 1.35, and 1.40. For the squares, the c values are (from the bottom): 1.20, 1.25, 1.30, 1.35, 1.40, and 1.45.

The data gives further support to "no halting". The number of trials needed to place n fill regions rises steeply with the number n of placements, but its trend is a power law which shows regular and predictable behavior and never goes to infinity. One test with discs made a million placements without halting.

The data in Fig. 7.5 has a persistent and interesting trend where the noise and scatter in the data are larger for larger c values.

We point out explicitly that the treatment of the halting problem given here is derived from computational experiments, and does not offer formal proof. The one-dimensional case is more tractable mathematically, and it is possible to show formally that there is a substantial range of c for which the algorithm is unconditionally non-halting (see Chapter 9).

Figure 7.5 is an extremely "busy" image and contains a large amount of very concentrated information. Careful study of it will greatly aid the reader's understanding.

Among other things, Fig. 7.5 can tell us on average how many trials are needed for a given run. Suppose that we wish to place 1000 discs with $c = 1.325$. We will use the vertical 10^3 line on the right-hand graph, keeping in mind that the logarithms of the quantities are what is plotted. The third data set from the bottom is for $c = 1.30$ and the fourth is for 1.35. So we will want to find the point on the 10^3 line halfway between these curves, which is about 5.4 as a base 10 logarithm. The average total number of trials needed is approximately $10^{5.4} \cong 250,000$.

It can be concluded from Fig. 7.4 that when a large number of shapes are to be placed with a high c value a huge number of trials is needed. This happens because the random search method is quite inefficient. Improved search algorithms may be possible and could substantially improve on this. An inefficient search can be tolerated with high-speed computers.

7.4. The "Black Area" and Halting

Figure 7.6 was constructed in the following way. To begin with, 25 disc positions were generated by the statistical geometry algorithm within a bounding circle. The bounding area was then colored black. A red disc with radius $r_i + r_{26}$ is then drawn at the location of each placed disc. A band of width r_{26} is then drawn around the perimeter of the bounding circle. The placed discs are then drawn in white. Any red areas that extend beyond the bounding circle are erased. Any random trial which puts the center of the next disc within any of the black regions will result in a successful placement of the next disc (disc 26). The total gasket area is the sum of the black area and the red area.

The black area is subject to two competing effects. As more discs are placed, they will erase some of the black area. However, those parts of the black area which did not get a placement will grow in size because the radius of the next-to-be-placed disc keeps going down.

It is evident that if one can show formally that the black area is always greater than 0, one has proof that the process never halts. This is the basis of the Ennis proof of non-halting [12] for discs.

The total black area A_b is closely related to the probability p_{pl} of a placement.

$$p_{pl} = A_b / A \tag{7.4}$$

where A is the area of the bounding circle.

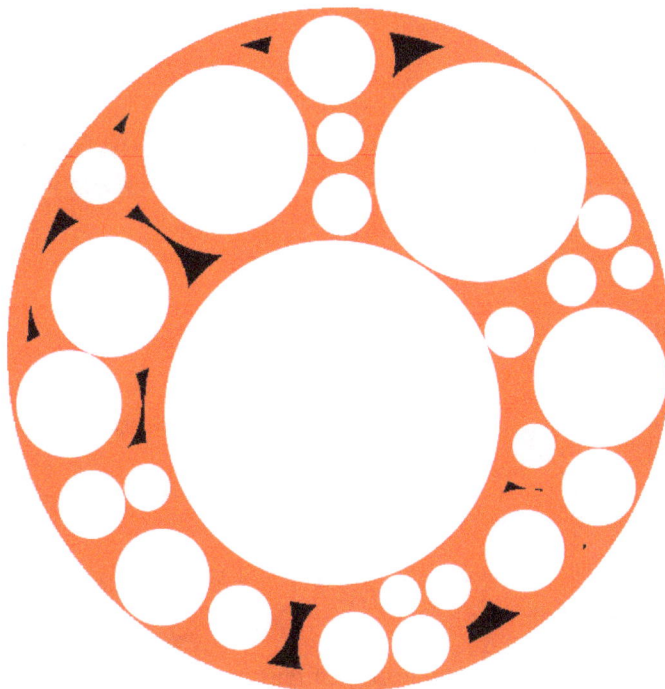

Fig. 7.6.   An example of the "black area" for discs fractalized within a circle.

The data in Fig. 7.5 is necessarily somewhat noisy because of the random nature of the placement algorithm, but a straight line with a slope +c is always a good approximation to the data when many discs have been placed. This was a puzzling observation for a long time. This behavior can be formulated mathematically as

$$n_{cum} = Bi^c \quad \text{when } i \to \infty \tag{7.5}$$

Here, B is a dimensionless proportionality coefficient and n_{cum} is the (average) number of trials needed to place i discs. It should be valid for large numbers of placements. From Eq. (7.5), we can pass to the placement probability P_{pl} by

$$P_{pl} = \left[\frac{dn_{cum}}{di} \right]^{-1} = \frac{1}{cBi^{(c-1)}} \quad \text{when } i \to \infty \tag{7.6}$$

Consider the equation

$$\frac{1}{\zeta(c)} \sum_{i=1}^{m} \frac{1}{i^c} + \frac{1}{m^{c-1}} = 1 + \theta(m) \tag{7.7}$$

Here, $\zeta(\ )$ is the Riemann zeta function (i.e., $N = 1$). It is found that $\theta(m)$ decreases monotonically and slowly to zero for large m. With $m = 100,000, \theta(m) = 0.002413$. If we assume $m \to \infty$, the two terms on the left-hand side must sum to 1. If the boundary has an area of 1, the first term is just the sum of the areas of the first m fill regions; thus the second term is the gasket area, which is seen to decrease by a power law with exponent $-(c-1)$. Thus for large i, we have the gasket area A_g:

$$A_g = G / i^{(c-1)} \quad \text{when } i \to \infty \tag{7.8}$$

Here, G is a proportionality coefficient which has the unit m². We can combine these equations to get

$$\frac{A_b}{A} = P_{pl} = \frac{1}{cBi^{(c-1)}} = \frac{A_g}{cBG} \tag{7.9}$$

This result says that the *average* black area (it has fluctuations) is a fixed fraction ($1/cBG$) of the total gasket area when i is large. This is supported by the trials versus placements data, and it accounts for the behavior seen in Fig. 7.5.

7.5. The Fractal Dimension

For those readers not familiar with the idea of fractals and fractal dimension, there are many sources of information [3]. The fractal dimension should not be confused with the physical or Euclidean dimension which has an integer value.

The fractal dimension D of a statistical geometry fractal in E Euclidean dimensions is given by

$$D = \frac{E}{c} \tag{7.10}$$

The fractal dimension does not depend on the details of the random numbers used and is the same for all fractals with the same c. The fractals shown in Fig. 4.2 of Chapter 4 all have the same fractal dimension.

Recursive geometric fractals such as Sierpinski have a single precisely defined value for the fractal dimension D. A unique feature of statistical geometry fractals is that fractal D can be any real number within the allowed range, i.e., it is a continuous variable rather than a fixed number.

Most accounts of geometric fractals treat recursive fractals, i.e., structures which are developed generation by generation. The Sierpinski fractal (Fig. 1.3 of Chapter 1) is a good example of this. There each new generation has three times more triangles than the previous generation, and each has a linear dimension which is half as large. If at each generation the number of shapes *increases* by a factor M and their linear dimensions *decrease* by a factor S, the fractal dimension for a recursive fractal is given by the following [3]:

$$D = \frac{\log(M)}{\log(S)} \tag{7.11}$$

For the Sierpinski triangles, this gives $D = \log(3) / \log(2) \cong 1.585$.

It is not easy to see how to apply this to non-recursive statistical geometry fractals. The approach taken here is to derive fractal D by analogy with Eq. (7.11).

We can develop the ideas of "number of shapes" and "length ratio" in a simple way which leads to Eq. (7.10). We think in terms of a continuous change in the size of the fill regions, rather than a discrete recursive one. Since the essence of a fractal is self-similarity, one can define "generations" here by *sampling* the properties of the fill regions at points equally spaced logarithmically.[a] Each successive sample will be viewed as playing the role of a generation in the recursive case.

Consider a two-dimensional statistical geometry fractal with $N = 1$. We can number the shapes $n = 1, 2, 3, \dots$. Suppose that we sample the fill region area distribution with samples taken for n values at

$$\ln(n_k + N) = k\Delta \tag{7.12}$$

Here, Δ is an arbitrary sample-spacing constant and the integer $k = 1, 2, 3, \dots$ numbers the samples. This relationship defines the n_k value at sample point k. If we exponentiate both sides of this equation, we find these sampling n values[b] to be

$$n_k + N = e^{k\Delta} \tag{7.13}$$

We define the "number of shapes" v_k in sample k to be

$$v_k = e^{(k+\frac{1}{2})\Delta} - e^{(k-\frac{1}{2})\Delta} = e^{k\Delta}(2\sinh(\Delta/2)) \tag{7.14}$$

Note that the minus sign in Eq. (7.14) causes N to cancel out. The ratio M of the numbers of shapes at samples k and $k + 1$ will be (independent of k)

$$M = \frac{n_{k+1} + N}{n_k + N} = \frac{e^{(k+1)\Delta}}{e^{k\Delta}} = e^{\Delta} \tag{7.15}$$

---

[a]In the Sierpinski case, one is also looking at points in the development (in length, number of copies) which are equally spaced logarithmically.

[b]The numbers n_k thus generated won't be integers as with Sierpinski. One could consider the region where n is quite large (after all the sequence is a power law everywhere) and argue that the n_k are "close enough" to integers. Since the power law is a continuous function, perhaps this is a non-issue.

Since area n is given by

$$A_n = (\text{const.}) \frac{1}{(n+N)^c} \tag{7.16}$$

then at sampling point k

$$A_k = (\text{const.}) \frac{1}{(e^{k\Delta})^c} = (\text{const.}) \frac{1}{e^{k\Delta c}} \tag{7.17}$$

So from one sample to an adjacent one, the *area* changes by a ratio ρ_A

$$\rho_A = e^{\Delta c} \tag{7.18}$$

while the *linear* size correspondingly changes by a ratio

$$S = e^{\Delta c/2} \tag{7.19}$$

From Eq. (7.11), we find

$$D = \frac{\log(M)}{\log(S)} = \frac{\log(e^{\Delta})}{\log(e^{\Delta c/2})} = \frac{\Delta}{\Delta c/2} = \frac{2}{c} \tag{7.20}$$

Note that Δ drops out, i.e., one gets the same result for any sampling interval. This approach to the fractal dimension shows how the idea of self-similarity can be understood in a situation without recursive generations. The answer agrees with Eq. (7.10), as it must. This procedure can be readily extended to the one-dimensional and three-dimensional cases by suitable changes. It follows that the fractal dimension does not depend on N, a conclusion which also follows from equations in the paper by Delaney *et al.* [6].

7.6. The Dimensionless Gasket Width

Since one is interested in determining whether the algorithm halts, it is of much interest to describe the average gasket width. Consider discs. If the average gasket width diminishes at the same rate as the diameters of the placed discs, one can expect that the algorithm will continue without halting. After much thought, the following definition was adopted:

$$W_{\text{ave}}^{\text{gasket}} = \frac{\text{gasket area after } i \text{ placements}}{\text{gasket perimeter after } i \text{ placements}} \tag{7.21}$$

The average gasket width will depend on c, N, and i. Figure 7.7 shows an example of a gasket. It is a narrow lacy object. $W_{\text{ave}}^{\text{gasket}}$ has the physical units of a length so it is a plausible candidate for the role of an average width. It is evident that $W_{\text{ave}}^{\text{gasket}}$ will decrease as more and more discs are placed. Equation (7.21) gives an explicit way of computing the average gasket width thus defined.

We will restrict the discussion to discs. What we are really interested in is the ratio of the average gasket width to the diameter of the next-to-be-placed disc. We define the *dimensionless* average gasket width after placement i, assuming a bounding region of unit area as follows:

$$\widehat{W}_{\text{ave}}^{\text{gasket}} = \frac{\text{gasket area after } i \text{ placements}}{(\text{gasket perimeter after } i \text{ placements})(\text{diameter of disc } i+1)} \tag{7.22}$$

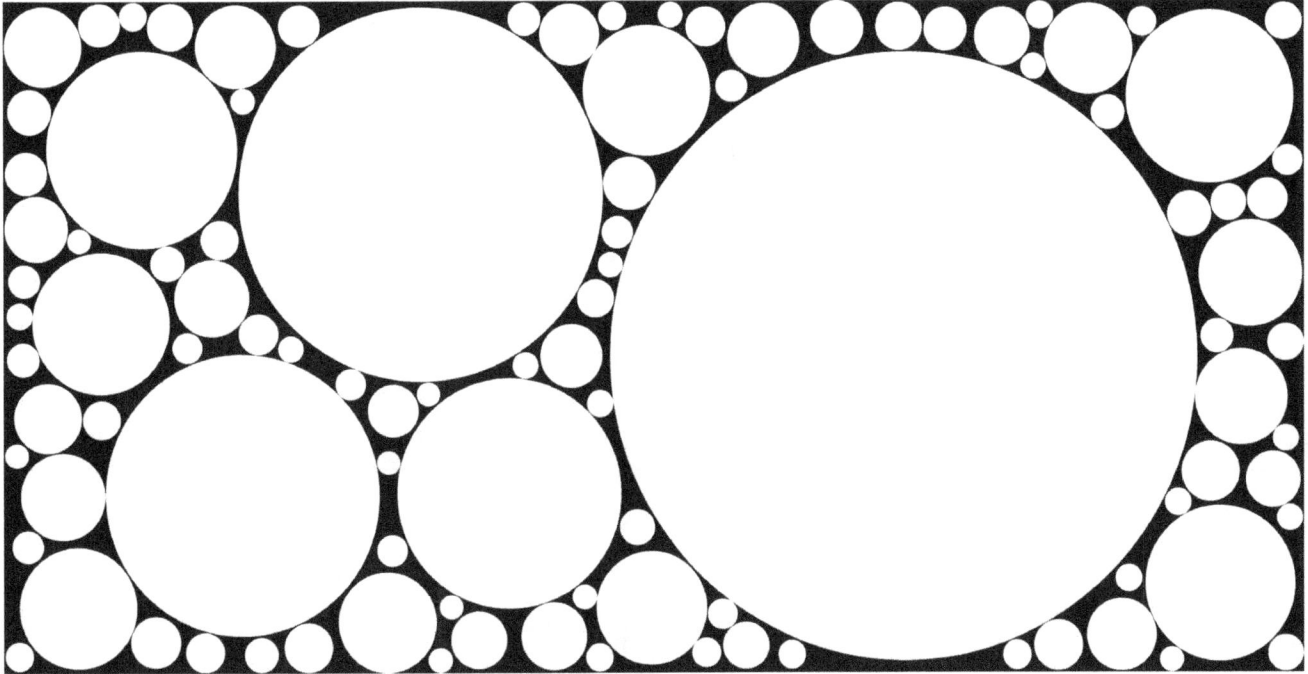

Fig. 7.7. An example of a gasket (black) for discs in a rectangle.

Thus, $\widehat{W}_{ave}^{gasket}$ is a dimensionless quantity, since the numerator has unit m^2, and the two factors in the denominator have units meter.

$$\widehat{W}_{ave}^{gasket} = \frac{1 - \dfrac{1}{\zeta(c,N)}\sum_{k=0}^{i}\dfrac{1}{(k+N)^c}}{2\sqrt{\dfrac{1}{\pi}\dfrac{1}{\zeta(c,N)(i+1+N)^c}}\,2\pi\sum_{k=0}^{i}\sqrt{\dfrac{1}{\pi}\dfrac{1}{\zeta(c,N)(k+N)^c}}} \quad (7.23)$$

This can be simplified somewhat to

$$\widehat{W}_{ave}^{gasket} = \frac{1}{4}\left[\frac{\zeta(c,N) - \sum_{k=0}^{i}\dfrac{1}{(k+N)^c}}{\sqrt{\dfrac{1}{(i+1+N)^c}}\,\sum_{k=0}^{i}\sqrt{\dfrac{1}{(k+N)^c}}}\right] \quad (7.24)$$

Figure 7.8 shows the behavior of $\widehat{W}_{ave}^{gasket}$ for discs. It can be seen that it has some variation for a small number of placements, but becomes nearly constant for many placements. The conclusion is that $\widehat{W}_{ave}^{gasket}$ falls in lockstep with the diameter of the next disc. This supports the idea that the algorithm does not halt. The limiting value of $\widehat{W}_{ave}^{gasket}$ for large n is small for large c, and large for small c.

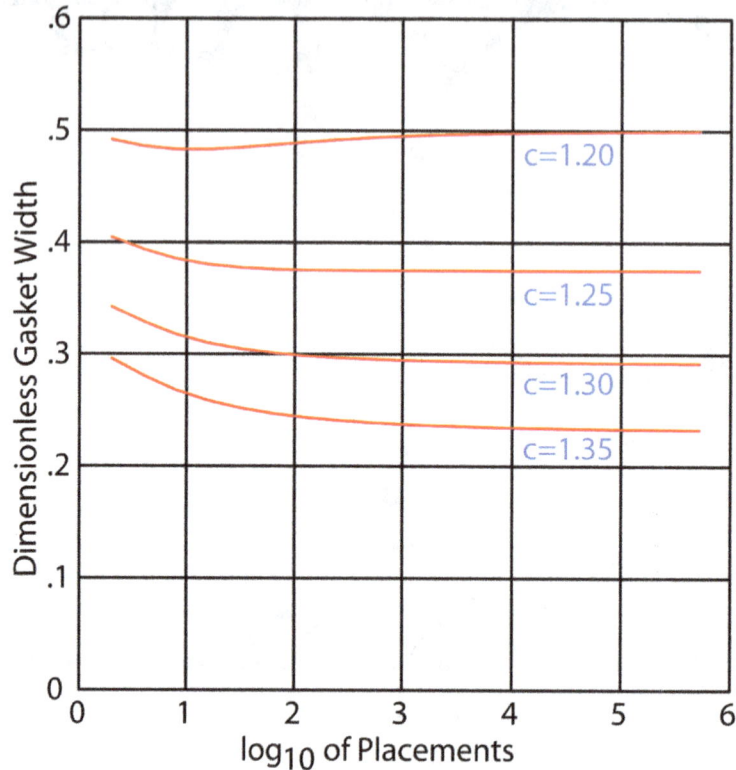

Fig. 7.8.　The dimensionless gasket width versus $\log_{10}$ of the number of placements, for four c values, when $N = 1$.

7.7. Order and Disorder

If we consider the variation of $\widehat{W}_{\text{ave}}^{\text{gasket}}$ with the exponent c, we see a quite substantial change. With small c, there is much more "wiggle room" than with large c. Examples of this in computed results can be seen in Figs. 2.5 and 2.9 of Chapter 2, Fig. 3.1 of Chapter 3, and Fig. 5.4 of Chapter 5. This goes far toward explaining the transition from randomness and disorder at low c to the close and orderly spacing of the shapes at high c (for example, Figs. 3.3 and 3.4 of Chapter 3).

We can identify c (or the corresponding fractal dimension D) as a parameter which governs order and disorder in these random fractals on the basis of the behavior of $\widehat{W}_{\text{ave}}^{\text{gasket}}$ and visually observed patterns. If the pattern gets more and more orderly as c increases (i.e., fractal D decreases), might the limit be the case where all of the fill regions are in a *completely ordered state*? What do we mean by this? Figure 7.9 illustrates the situation for triangles.

Edge effects are minimal when triangles are fractalized within a triangle, so Fig. 7.9 is a cleaner example than rectangular boundaries (Fig. 3.1 of Chapter 3). One can see a steady progression from randomness to short-range order in the sequence in Figs. 7.9(a)–7.9(d). Short-range order[c] in Fig. 7.9(d) takes the form of triplets of small triangles near a given triangle and is most pronounced for the smallest triangles. These triplets resemble those of Sierpinski (Fig. 7.9(e)) in both relative size and neighbor configuration, but with a certain amount of "noise" due to the random process involved. Some of this short-range order can be seen in Fig. 7.9(c), but it is not evident in Figs. 7.9(a) and 7.9(b). As c goes from 1.12 to 1.24, there is a steady drop in the fractal dimension D, which tracks with c. We can

---

[c]Range of order is an established concept in statistical physics. Two objects (e.g., atoms) are said to be ordered if the position of one has a degree of correlation with the position of the other. If nearby objects are highly correlated but distant ones have little correlation, there is said to be short-range order (as in liquids). If objects are highly correlated with other objects even to great distances (as in crystals), there is said to be long-range order.

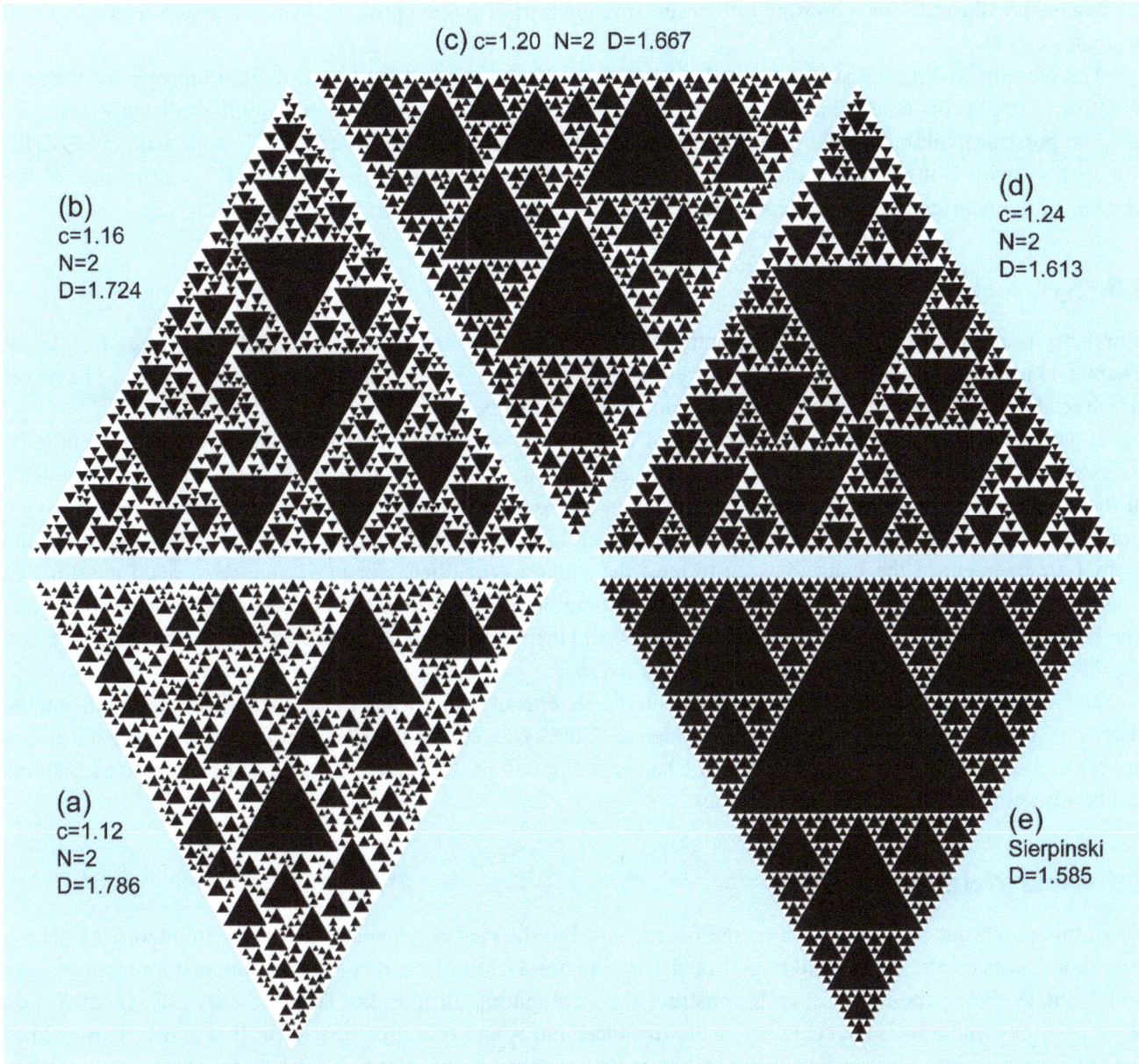

Fig. 7.9. Triangles fractalized within a triangle (like Fig. 3.2 of Chapter 3), with a progression of c values from the smallest (a) to the largest (d). There is a chaos-to-order progression from (a) to (d), with Sierpinski (e) representing perfect order. The smallest triangle has linear dimensions which are 4% of the largest in (a)–(d).

view D as controlling order versus disorder. High D (near 2) corresponds to randomness, while low D corresponds to correlation and order. It is interesting that D for Sierpinski (Fig. 7.9(e)) is less than 2% lower than with $c = 1.24$ (Fig. 7.9(d)). One can view Sierpinski as the ultimate limit of a highly-ordered triangle fractal. From this viewpoint, it imposes a limit on the highest usable value of c, namely $2/1.585 = 1.262$, which agrees with observation. We can summarize the situation thus:

For low c (Fig. 7.9(a)), there is neither short-range nor long-range order.
For high c (Fig. 7.9(d)), there is short-range but no long-range order.
For Sierpinski (Fig. 7.9(e)), there is both short-range and long-range order.

Figure 7.9 illustrates the opposing influences of random trials (which promote chaos) and non-overlap (which promotes order).

The patterns in Fig. 7.9 suggest that there is a form of entropy involved, with higher entropy for lower c. What form might this take? Shannon entropy doesn't fit well. Thermodynamic entropy doesn't work since we have no pressure, volume, or temperature. The most interesting candidate for entropy S is perhaps $S = \log(W)$ (by analogy with Boltzmann's famous result $S = k \log(W)$ in statistical physics) where W is a measure of the number of geometric states possible for the system.

7.8. Symmetry

Symmetry is an interesting aspect of geometry [1]. One of the charming features of tessellations (e.g., Fig. 1.1 of Chapter 1) is their many elements of symmetry: translation vectors, rotation centers, mirror planes, etc. The recursive fractal of Fig. 1.3 in Chapter 1 also has symmetrical elements.

If we look at the random disc fractal of Fig. 4.1 in Chapter 4, we see that while it appears to be quite orderly, *it has no symmetry* of the usual kinds (rotation, translation, mirror, etc.). Its only formal symmetry is the symmetry of the disc itself. But to many viewers, such a pattern seems to have a certain regularity which we do not have a simple word for. A technical term for the regularity seen in Fig. 4.1 of Chapter 4 is *statistical self-similarity*; the mutual arrangement of the shapes is seen to have the same *average* behavior *at all length scales*. Figure 6.1 of Chapter 6 is another good illustration of statistical self-similarity. The use of log-periodic color helps to bring this out. If you focus on one color in Fig. 4.1 of Chapter 4 and then another (i.e., one size of shape versus another), you see that the mutual arrangement is the same *on the average*.

The symmetry of the shapes does play a role in the patterns of filling when c approaches its upper limit and the shapes have little "wiggle room" [13]. The influence of the symmetry of the shape on filling patterns is best seen in Figs. 3.3–3.5 of Chapter 3, and Fig. 5.3 of Chapter 5. We will see in Sec. 7.10 that statistical geometry patterns can be made to have space-group symmetries.

7.9. Near Neighbors

Up to this point, our focus has been on the fill regions, but the spaces between them are also interesting. These neighbor spaces can be quite small (e.g., Fig. 4.1 in Chapter 4). For discs, it is easy to construct a *proximity map* which shows their pattern. The map is constructed by computing all interdisc (edge-to-edge) distances d, i.e., $d = r_{12} - r_1 - r_2$ where r_{12} is the center-to-center distance, and r_1 and r_2 are the disc radii. If d is less than g times the smallest (i.e., last-placed) disc radius, a line is drawn between the centers of the discs, otherwise no line is drawn. Figure 7.10 shows such a construction.

The discs which are linked in the proximity map are shown in light gray, while those which have no links are shown in blue. A diagram like Fig. 7.5 is a *planar graph*.[d] In the language of graphs, the center-to-center lines are edges and the centers of the linked discs are vertices.

We see that there are linked subgraphs of all sizes from a single edge to quite complicated ones. The appearance of these graphs varies greatly depending on the particular values of c, N, n, and g. As $g \to 0$, the graph has only a small number of edges distributed over subgraphs which seldom have more than two or three vertices. As $g \to 1$, the graph becomes a single entity with all of the disc centers linked. The intermediate cases are the most interesting and varied. The strong constraint of prior placements which influences disc placement also affects proximity maps. This influence results in graphs which have a high degree of randomness, but are far from purely random.

---

[d]If g becomes large enough, some of the lines can make crossovers to second-level neighbors and the graph will no longer be planar.

Fig. 7.10.   A proximity map for discs. $c = 1.42$, $N = 80$, and $g = 0.25$. There are 1500 discs, 1196 vertices, and 1015 edges.

Ordinarily, if the list of vertices and edges of one graph can be made identical to the list for another by renumbering, the graphs are thought of as the same. Because of the geometric construction of these graphs, the lengths of the edges and the edge angles at vertices have physical meaning so that two graphs which are the same in the ordinary sense are not the same from physical and geometric considerations.

Such proximity maps can in principle be constructed for any shape but the computations are easier for discs. When spheres are placed in three dimension, a proximity map for them would be a non-planar graph.

Fig. 7.11.   A fractal with a hollow shape colored by hierarchical rank.

7.10. Hollow Shapes and Hierarchy

If the shape is hollow (like the annular rings of Fig. 4.3 in Chapter 4), the gasket is no longer a single connected whole, but breaks up into a large number of pieces defined by the interiors of the shapes. It is possible to arrange the shapes hierarchically with *ranks* labeled by an integer as shown in Fig. 7.11. Each rank is assigned a unique color.

This hollow fill region has a circle for its exterior and a pair of circular arcs which define its interior. It somewhat resembles an open mouth. Fill regions without anything in them have rank 0; those with only rank 0 objects inside them have rank 1, etc. There is 1 shape of rank 3, 13 of rank 2, 36 of rank 1, and 101 of rank 0. In general, if the highest ranked shape within a given shape is rank m, its rank is $m + 1$. Rank tracks closely with size, and changes at size boundaries. Sometimes people see themselves as occupying a place in a random hierarchy like this.

Rank plays somewhat the same role as generation number in recursive fractals (e.g., Sierpinski, Fig. 1.3 of Chapter 3). When a log–log plot is made of the number of rank m shapes versus the mean linear dimension of the rank m shapes, the data gives a good fit to a straight line.

7.11. Tessellations

A simple tessellation based on rectangular periodic boundaries is shown in Fig. 2.7 of Chapter 2. This form of tessellation has translation symmetry, but does not have any mirror lines, etc. and is thus limited. Can one impose mirror symmetry within the bounding region which can be tiled? Consider discs; mirror symmetry would require that none of the discs can touch the boundary ("inclusive boundary"), or that all of the boundary discs must be placed with their centers precisely on the boundary.

The latter case can be achieved proceeding as usual for periodic boundaries, but *when any disc intersects the boundary it is moved perpendicular to the boundary until its center is precisely on the boundary* before doing the overlap test. Adjustment must be made for the fact that only half the area of such a boundary disc is within

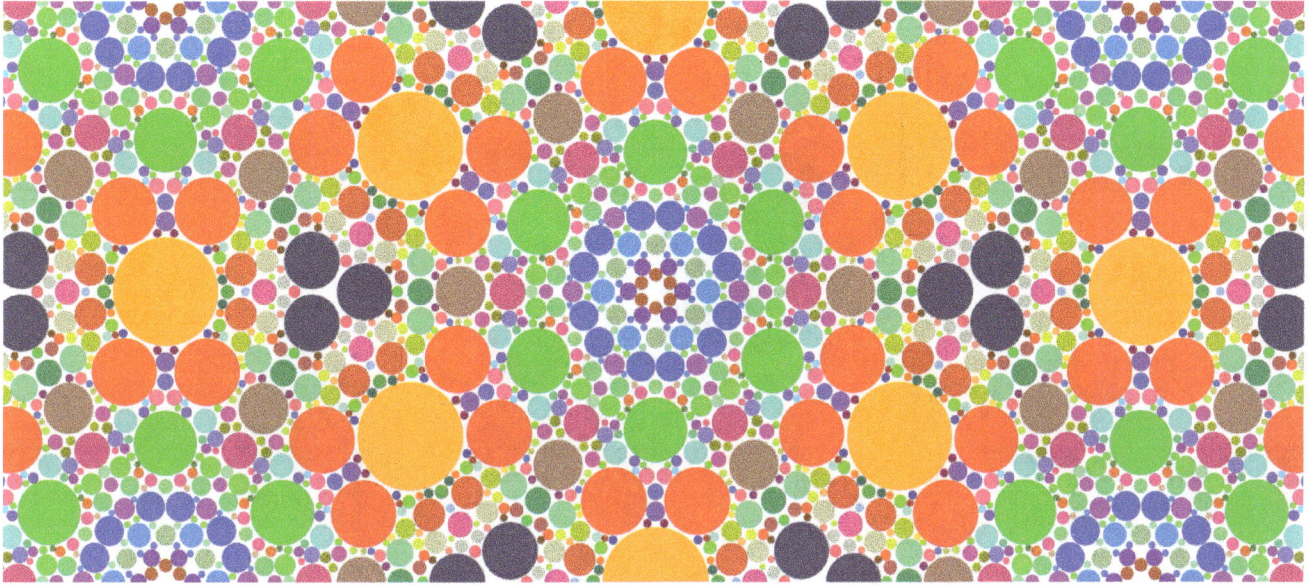

Fig. 7.12.   A p6mm tessellation with boundary discs whose centers are on the boundary. Modified random color. Equivalent discs all have the same color.

the bounding region if the area rule is to be satisfied. This is done by doubling the area of the disc when it is placed on the boundary. An example with p6mm symmetry [1] is shown in Fig. 7.12.

The pattern in Fig. 7.12 has a conflict between two opposing tendencies. The p6mm symmetry tends to make it look orderly, while the randomness within the unit tile contributes noisy disorder. The color scheme used greatly eases the problem of seeing the symmetry, but if the pattern were done in a single color, the eye would perceive mostly the randomness. Many would call such a tiling "kaleidoscopic".

7.12. Finite Spacing

In Fig. 7.9 (triangles), we see that when c is small (approaching 1) there is much "free space" in the arrangement. Can one take advantage of this to impose a constraint that causes discs to "repel" each other and assume a more orderly arrangement? The answer is yes, and Fig. 7.13 shows an example.

In this scheme, the area rule is followed as usual, but during the random search a location which can accommodate a disc of radius β times ($\beta > 1$) the area-rule radius is sought. A disc with the area-rule radius is then placed in the database. Since the usual area rule is followed, the modified algorithm remains space-filling. The number of trials needed for a placement can be much larger in such a scheme, and the maximum c is correspondingly lower. Available evidence indicates that within the allowed range of c, the algorithm is non-halting. In Fig. 7.13, the β value of 2 ensures that every perimeter-to-perimeter spacing is at least the radius of the smallest disc. The visual impression of such a pattern is quite different than others shown in the book, with the "grouted" separation between discs making each disc stand out.

It is also possible to get various degrees of correlation by manipulating β with grouped shapes. Figure 7.14 shows the results when we separate the discs into groups, with "self" and "other" β values different. The arrangement on the right has the different disc groups concentrated in random regions, with some resemblance to Fig. 5.3 in Chapter 5. The view taken here is that the pattern on the right is correlated, while the one on the left is anticorrelated.

Fig. 7.13.   "Mutually repelling" discs in a circle with $c = 1.15$, $N = 3$, 1000 discs, and $\beta = 2$.

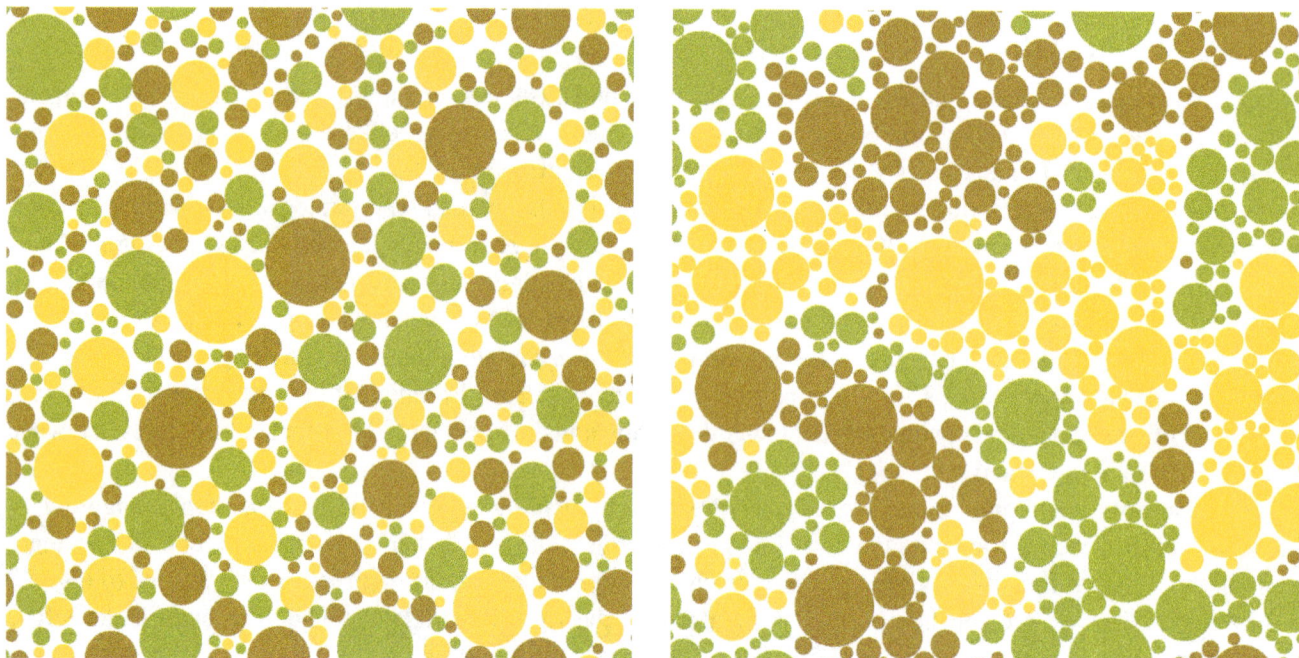

Fig. 7.14.   Grouped discs. The sets of discs are 0, 3, 6, …; 1, 4, 7, …; and 2, 5, 8 … with each set having its own color. On the left, we have $\beta_{00} = \beta_{11} = \beta_{22} = 1.5$ and $\beta_{01} = \beta_{02} = \beta_{12} = 1.0$. On the right, the two values 1.5 and 1.0 are interchanged. $c = 1.31$, $N = 20$, 500 discs, and 64% fill.

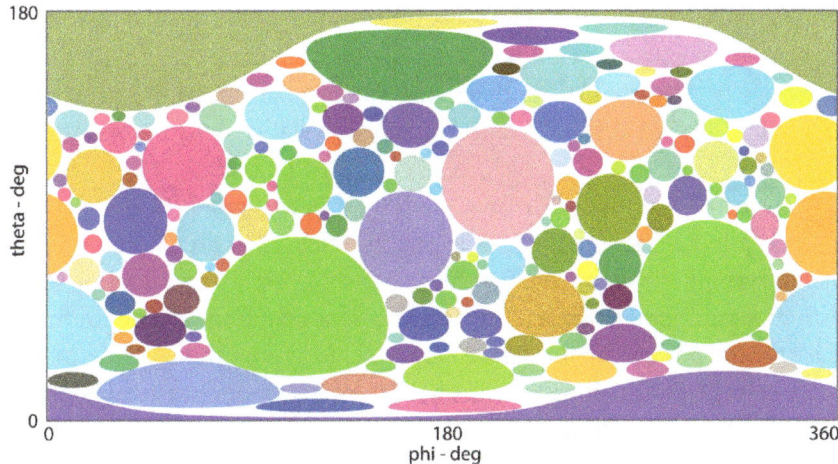

Fig. 7.15.   A projection of spherical caps fractalized on the surface of a sphere.

7.13. Spherical Surfaces

Can one fractalize spherical caps on a spherical surface? The answer is yes, as shown in Fig. 7.15. This image is a projection of the sphere and the caps using the spherical coordinates θ and ϕ. Such a projection has much distortion near the poles. The caps appear nearly circular near the equator, but greatly elongated near the poles. The green and purple regions at the top and bottom show spherical caps that contain the north and south poles. On a spherical surface, there is no distinction between inclusive and periodic boundaries.

7.14. Problems and Questions

The algorithm is expressed in purely mathematical terms, and it can be studied by the methods of research mathematics. There are a number of problems and conjectures which invite further study:

(1) While a start has been made on formal non-halting proofs, much remains to be done.
(2) Is it possible to find explicit expressions for c_1 and c_2 in Fig. 7.3?
(3) Can an entropy be defined for the order versus disorder?
(4) Can maximum c be predicted from the shape of the fill region?
(5) What is the asymptotic form of the dimensionless gasket width as $i \to \infty$?

7.15. Power-Law Distributions

Many natural and man-made phenomena have been found which obey power-law statistics [14, 15]. Those who study such phenomena say that a statistical distribution is "fractal" if it obeys a power law. One of the best documented natural examples is the Gutenberg–Richter law [14] for earthquake energy release. It says that the number of earthquakes with energy release E is proportional to E^{-2}. (In analyzing data from seismometers, the exponent is not exactly 2, but a value like 1.98 or 2.01.) Thus earthquakes with large energy release are much rarer than those with small energy release. The energy release E has a huge range of values, from the magnitude 9 earthquakes which can kill thousands (and only happen a few times per century) to the smallest ones detectible with seismometers which form a sort of continuous noise.

An example of a man-made power-law distribution is the sizes of US businesses ranked by number of employees. Another one is the distribution of cities by population. In both cases, there are lots of little ones and only a few big ones. Several hundred such power-law distributions are known. They are quite different than the Gaussian and other distributions found in statistics textbooks. The probability of the textbook distributions falls off from the mean as $exp(-x)$ or $exp(-x^2)$. A power law x^{-v} with exponent v falls off much more slowly than that, so that such distributions are sometimes called "fat-tailed". The mean and standard deviation are meaningless for such distributions. Such statistical distributions are largely absent from college statistics textbooks although many of them are known.

Another feature of power-law distributions is that the largest events completely dominate. The energy release in the largest earthquakes is a large fraction of the total energy released in all earthquakes. This can be seen in all of the statistical geometry examples. The largest few fill regions take up a huge fraction of the total area.

One of the interesting possibilities is the use of the statistical geometry algorithm for modeling natural phenomena. A difference is that the power law for earthquakes, etc. is approximate, while for the statistical geometry fractals, it is exact. These natural distributions have a smallest event, while statistical geometry distributions allow the fill region areas to diminish without limit.

Chapter 8

The Area–Perimeter Algorithm

8.1. Introduction

In this algorithm, the area rule of Chapter 7 (Eq. (7.1)) is replaced by a different rule which is recursive and at first sight seems odd. We specifically consider discs.

$$r_{i+1} = \gamma \frac{A_i}{P_i} \tag{8.1}$$

$$A_{i+1} = A_i - \pi r_i^2 \tag{8.2}$$

$$P_{i+1} = P_i + 2\pi r_i \tag{8.3}$$

Here, r_{i+1} is the radius of disc $i+1$, A_i is the gasket area after placement i, and P_i is the corresponding gasket perimeter. Figure 7.7 of Chapter 7 illustrates a gasket — the lacy black area between placed fill regions. Placement of the fill regions is done by the same rule as for the statistical geometry algorithm (Fig. 7.1 of Chapter 7).

Rather than an area rule, this is a radius rule. In statistical geometry, the areas of the fill regions are computed using the prescribed area rule. In this algorithm, they are computed recursively from the results of the previous step. Since the radius is simply related to the area, this can also be viewed as an area rule.

In Sec. 7.6 of Chapter 7, it was found that the dimensionless gasket width $\widehat{W}_{\text{ave}}^{\text{gasket}}$ goes to a constant value after many fill regions have been placed. This was the inspiration for Eq. (8.1), for which $\widehat{W}_{\text{ave}}^{\text{gasket}}$ is precisely constant. We can rearrange Eq. (8.1) to

$$\frac{1}{2\gamma} = \frac{A_i}{2r_{i+1}P_i} \tag{8.4}$$

It will be recognized that the right-hand side of Eq. (8.4) is $\widehat{W}_{\text{ave}}^{\text{gasket}}$, and that it has the fixed value $1/2\gamma$. Since both the statistical geometry algorithm and the area–perimeter algorithm have a constant dimensionless gasket width for many fill regions placed, we can expect similar behavior for both algorithms as $i \to \infty$.

8.2. Computed Results

Equations (8.1)–(8.3) replace the area rule of statistical geometry. The value of γ is chosen *a priori* and remains constant throughout the process. When this is done for discs in a circle, we get a result like Fig. 8.1. The first few

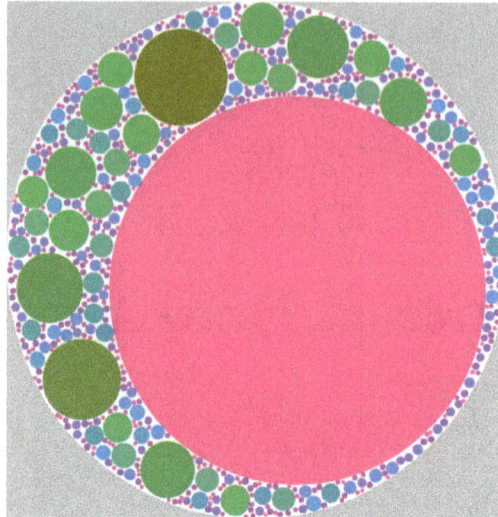

Fig. 8.1.   Placement of 600 discs within a bounding circle using the area–perimeter algorithm, $\gamma = 3/2$. Log-periodic color.

iterations can be thought of as a *launching* process. There is some ambiguity as to the initial value of the perimeter. In Fig. 8.1, the initial perimeter was taken as the perimeter of the bounding circle. Under this assumption, γ cannot exceed 2. Higher γ values can be used with different launch schemes. The algorithm runs satisfactorily if an initial sequence of radii not obeying Eq. (8.1) makes a transition to the sequence of Eq. (8.1) at some point. One way this can be done is to compute two sequences of disc radii and areas, taking the lower radius value at each point such that Eq. (8.1) applies for large i.

The pattern shown in Fig. 8.1 resembles typical statistical geometry fractals except for the very large disc 1.

If the bounding circle has radius R, the area and perimeter for the bounding circle are πR^2 and $2\pi R$, respectively, so that the initial area/perimeter ratio is $R/2$. If $\gamma > 2$, the first disc does not fit within the bounding circle. In Fig. 8.1, it can be seen that if the largest disc is placed near the center of the bounding circle, there is not enough room for the second disc. If the algorithm does not halt at the placement of disc 2, it continues without halting in computer runs with $\gamma = 3/2$.

One of the most interesting questions is "Does the algorithm halt?" (if the placement of the second disc is successful). Figure 8.2 shows a log–log plot of the cumulative number of random trials needed to place n discs for Fig. 8.1. Because of the random numbers used, such a plot has noise in it, but it can be seen that a straight line is an excellent regression for the data beyond about 100 placements. Thus, the data of Fig. 8.2 indicates that this run would not halt but continue indefinitely.

The sequence of disc radii generated from Eq. (8.1) can be computed without reference to the random numbers used in the trials. It is particularly interesting to make a log–log plot of the disc areas versus placement number to see if this relationship is a power law. The exponent of such a power law can be used to compute the fractal dimension via Eq. (7.10) in Chapter 7. Because such an exponent plays the same role as c in Chapter 7, we use the same symbol for it.

Such a log–log plot is shown in Fig. 8.3. The red dots are the disc areas, plotted for placement numbers that are powers of 2. With this choice, the dots are equally spaced along the X-axis. The log slope was calculated for each adjacent pair of points. Tests of convergence showed that the calculated log slope (i.e., the exponent c) was accurate to about six significant decimals for the last two points. Because the results were so surprising, we give the raw data in Table 8.1.

gamma=1.50

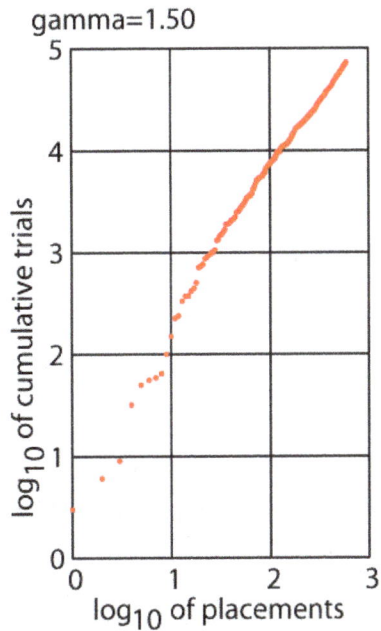

Fig. 8.2.   Log–log plot of the cumulative number of random trials needed to place n discs with $\gamma = 3/2$.

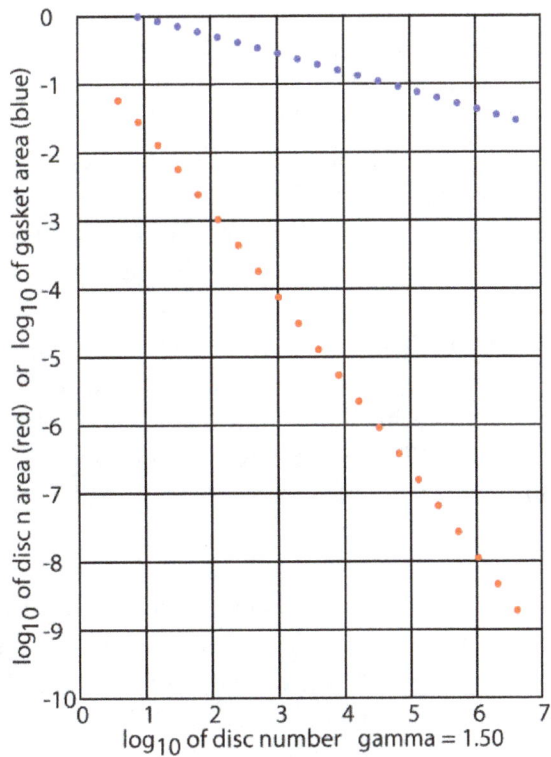

Fig. 8.3.   Log–log plots of the disc areas (red) and the gasket area (blue) for $\gamma = 3/2$.

Table 8.1.   Computed exponent c values with the corresponding fractal dimension D, versus γ for discs.

γ	c			$D = 2/c$
1.0	1.20000	(6/5)	[12/10]	1.66667
1.5	1.27272	(14/11)	[14/11]	1.57143
2.0	1.33333	(4/3)	[16/12]	1.50000
2.5	1.38461	(18/13)	[18/13]	1.44444
3.0	1.42857	(10/7)	[20/14]	1.40000

In the initial work, γ had values of 0.5, 1.0, and 1.5. What was quite surprising was the appearance of repeating or terminating decimals for the log slope. Further exploration showed that this persisted for all of the cases studied. Is there a simple formula for c as a function of γ? The conclusion from the numerical data was that if one uses a quotient of two integers for γ, one gets a quotient of two other integers for c. The integer fractions shown in square brackets in the second column of Table 8.1 have a quite regular progression which obeys the following equation:

$$c = \frac{4 + 2\gamma}{4 + \gamma} \tag{8.5}$$

It is not difficult to see that Eq. (8.5) fits all the data with a precision of about six decimal places. This is a surprising result. If $\gamma > 0$, the c value from Eq. (8.5) is always greater than 1. For very large γ values c approaches 2, but computation shows that the algorithm often halts for γ greater than approximately 3 (i.e., a c value of 1.42857). This is in line with the highest c values usable for discs with the statistical geometry algorithm. The apparent occurrence of integers as coefficients in Eq. (8.5) is a startling feature.

Equation (8.5) should follow directly from Eqs. (8.1)–(8.3) when $i \to \infty$. The random numbers used in the trials do not affect the results in Table 8.1. In principle, this is just another calculus limit problem, but it has so far defied solution.

The statistical geometry algorithm of Chapter 7 is by construction space-filling in the limit. Is this algorithm also space-filling? The blue dots in Fig. 8.3 show the gasket area. This is another negative-exponent power law. The numerical calculations show that this exponent is $-(c-1)$ to six decimal places. (Is there another relationship similar to Eq. (7.7) in Chapter 7 involved here?) This power law for the gasket area indicates that one can make the gasket area arbitrarily small with enough placements, i.e., this algorithm is also space-filling in the limit.

For Fig. 8.2, the log slope of the trials versus placements data when more than 100 discs have been placed is about +1.27 (i.e., it has the value $-c$). This is what we would expect from the discussion in Sec. 7.4.

The initial stage of the algorithm has substantial flexibility. The algorithm stated above cannot run if $\gamma > 2$ because the first disc will not fit. Another approach is to use a different launch scheme with discs of fixed size, and at every step compare the fixed radius with the radius given by Eq. (8.1), taking the lower value for the next radius. Figure 8.4 shows the result with $\gamma = 5/2$. The numerically computed exponent c is -1.384615 (18/13), which agrees with Eq. (8.2) in spite of the different method used for the first few discs. The percentage fill is much higher than in Fig. 8.1.

Fig. 8.4.   400 discs placed within a bounding circle with $\gamma = 5/2$. The radius of the early fixed-size discs (orange) is 0.18 times the radius of the bounding circle.

8.3. Higher Dimensions

If we consider spheres in three dimensions, the equations become:

$$r_{i+1} = \gamma \frac{V_i}{A_i} \tag{8.6}$$

$$V_{i+1} = V_i - \frac{4}{3}\pi r_i^3 \tag{8.7}$$

$$A_{i+1} = A_i + 4\pi r_i^2 \tag{8.8}$$

Here, V_i is the gasket volume and A_i is the gasket surface area. As previously, γ is a dimensionless parameter. In the quotient on the right-hand side of Eq. (8.6), we divide a quantity having unit m³ by one having unit m², so that the quantity on the right-hand side has units meter.

In more generality, we can define V_i and A_i as the volume and area of the gasket in E Euclidean dimensions. A disc is to be thought of as a 2-sphere, a sphere is to be thought of as a 3-sphere, etc. The values of α and β in Eqs. (8.10) and (8.11) for the four-dimensional shapes are taken from the Wikipedia articles on "*N*-sphere" and "*N*-cube".

Table 8.2.   Parameters and formulas.

Shape	E	α	β	β/α	c formula
2-sphere	2	π	2π	2	$c = (4 + 2\gamma)/(4 + \gamma)$
3-sphere	3	$(4/3)\pi$	4π	3	$c = (9 + 3\gamma)/(9 + 2\gamma)$
4-sphere	4	$(1/2)\pi^2$	$2\pi^2$	4	$c = (16 + 4\gamma)/(16 + 3\gamma)$
2-cube	2	1	4	4	$c = (8 + 2\gamma)/(8 + \gamma)$
3-cube	3	1	6	6	$c = (18 + 3\gamma)/(18 + 2\gamma)$
4-cube	4	1	8	8	$c = (32 + 4\gamma)/(32 + 3\gamma)$

Let there be parameter sequences u_i, V_i, and A_i defined by the recursion

$$u_{i+1} = \gamma \frac{V_i}{A_i} \tag{8.9}$$

$$V_{i+1} = V_i - \alpha u_i^E \tag{8.10}$$

$$A_{i+1} = A_i + \beta u_i^{E-1} \tag{8.11}$$

Because this is fundamentally about geometry, it is desirable to be specific about the physical units involved. The u parameter has dimensions of a length (meter). The quantities in Eq. (8.10) have unit m^E, and those in Eq. (8.11) have unit m^{E-1}. The parameters γ, α, and β are dimensionless positive real numbers, and E is a positive integer. Careful numerical work shows that for large i, the generalized volume (αu_i^E) goes over to a negative-exponent power law versus the number of iterations i as in Fig. 8.3. We define the exponent of this power law to be c. K is a constant of proportionality.

$$\alpha u_i^E \to K i^{-c} \quad \text{as } i \to \infty \tag{8.12}$$

We have studied two families of fill region shapes (cubes and spheres) in two, three, and four dimensions. The relevant parameters are given in Table 8.2. For the sphere family, u is the radius, while for the cube family, it is the edge length. The agreement between the formulas given in Table 8.2 and numerical computations of c extends to between five and six significant digits. An example of the (rather short) code used in these computations can be found in Sec. 10.6 of Chapter 10.

8.4. Discussion

The reader will see that the formulas for c have integers in them. One of the surprising results of the early work on discs was that if one chooses a quotient of integers for γ, one gets a different quotient of integers for c. Similar features are seen here. The formulas of Table 8.2 give an excellent fit with the computed c values.

The results for the N-sphere family can be put in the general form as follows:

$$c = \frac{E^2 + E\gamma}{E^2 + (E-1)\gamma} \tag{8.13}$$

Here, E is the Euclidean dimension. This is another surprising result. Integers and their squares appear. A number of conjectures are posed:

(a) The coefficients in the c formulas are integers.
(b) The coefficient of γ in the numerator of the c formula is E.
(c) The coefficient of γ in the denominator of the c formula is $E-1$.
(d) The constant term in Eq. (8.13) is E^2.
(e) Why do the c formulas have the quotient form in all cases?

Equation (8.13) seems to be trying to tell us something fairly significant about Euclidean space but what is it? Equation (8.5) can be solved for γ, with the result

$$\gamma = \frac{4(c-1)}{2-c} \tag{8.14}$$

Since the dimensionless gasket width $\widehat{W}_{ave}^{gasket}$ is $1/2\gamma$ (see Eq. (8.4)) we can use this to find that

$$\widehat{W}_{ave}^{gasket} = \frac{1}{8}\left[\frac{2-c}{(c-1)}\right] \tag{8.15}$$

When c is only slightly higher than 1, we would expect $\widehat{W}_{ave}^{gasket}$ to be large, and that is what happens with Eq. (8.15) because of the $(c-1)$ factor in the denominator.

Equation (8.15) should also be valid for statistical geometry when $i \to \infty$. If we calculate $\widehat{W}_{ave}^{gasket}$ for $c = 1.2$ from Eq. (8.15), we find a value 0.5, in excellent agreement with the computed data in Fig. 7.3 of Chapter 7 for large i.

If $\gamma = 1$, for the 2-sphere we find $c = 6/5 = 1.20$; for the 3-sphere we find $c = 12/11 = 1.09$; for the 4-sphere we find $c = 20/19 = 1.05$. This agrees with the general trend that the highest usable c value drops rapidly in higher Euclidean dimensions.

Chapter 9

Statistical Geometry in One Dimension

While statistical geometry fractals in one Euclidean dimension do not produce the intricate images seen with two dimensions, their mathematical study is more tractable. This chapter begins with an empirical description of these fractals similar to the two-dimensional case (Chapter 7). It is possible to make much more detailed studies of the one-dimensional algorithm as a stochastic process than are possible in two dimensions and this is taken up in Sec. 9.6. The one-dimensional form of the dimensionless average gasket width is studied in Sec. 9.7 and it is demonstrated that it provides a tool for studies of the halting process leading to the conclusion that the one-dimensional algorithm with single segments is unconditionally non-halting when $1 < c < 2$ and $N = 1$.

9.1. General Description

The algorithm in one dimension is the same as the two-dimensional case described in Sec. 7.1 of Chapter 7, where "area" is replaced by "segment length". Random trials are used to find non-overlapping places for smaller and smaller line segments. The larger line segment in which the fractalized segments are placed is called the "container segment". A bar chart is a good way to illustrate such a fractal in one dimension (Fig. 9.1).

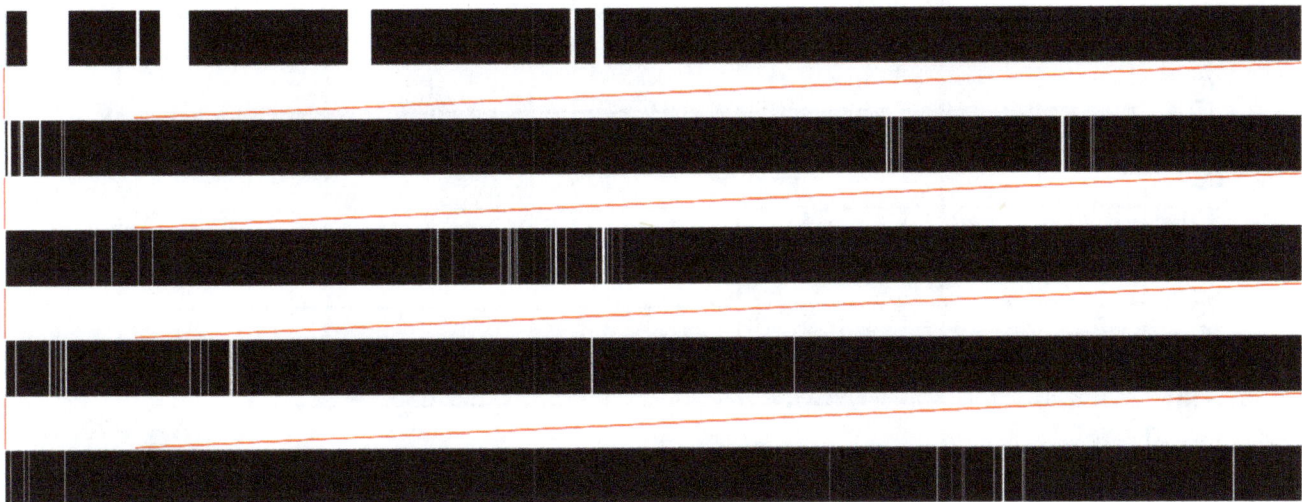

Fig. 9.1.   Bar chart of a one-dimensional random fractal with $c = 1.90$, $N = 1,998$ placed segments, and 99.87% fill. The bottom bar shows the complete fractal, while those above show progressive 10:1 expansions of the left end. The fine detail in the lower bars is not resolved.

The one-dimensional fractal can be thought of as a set of black bars (the placed segments) with a large number of ever-smaller white inter-segment regions (the gasket) sliced through it. One significant difference is that it is possible to achieve extremely high percentage fills in the one-dimensional case. Another difference is that quite high c values can be used. In two dimensions, $c = 1.50$ was about the best that could be done (with $N = 1$). In one dimension, it is possible to use much higher c values. Because of the very fine detail in these fractals, all of the one-dimensional computations have used double precision arithmetic.

9.2. Halting Data

Figure 9.2 shows computed halting behavior for a one-dimensional single-segment fractal when $N = 1$. Each data point is the average of 1500 runs. The trend follows the general scheme of Fig. 7.2 in Chapter 7. It can be seen that c_1 is around 2, while c_2 is well above 3.5. Data on the placement number at which halting occurs is qualitatively similar to that for the two-dimensional case [8] and shows the same pattern of "infant mortality" in the halting statistics. There is some scatter in the data because of the finite number of runs made.

9.3. Run-Time Behavior

Figure 9.3 shows a log–log plot of needed trials versus placed segments. The overall pattern is the same as in Fig. 7.5 of Chapter 7, with the data lying along straight lines for large n. Thus, the trend in cumulative trials versus placements is a power law, as was seen previously in the two-dimensional data (Fig. 7.5 of Chapter 7). The log slopes of the data lines in Fig. 9.3 are about the same as the corresponding c values. Because of these steep slopes, the number of trials needed to place segment n grows very steeply with n.

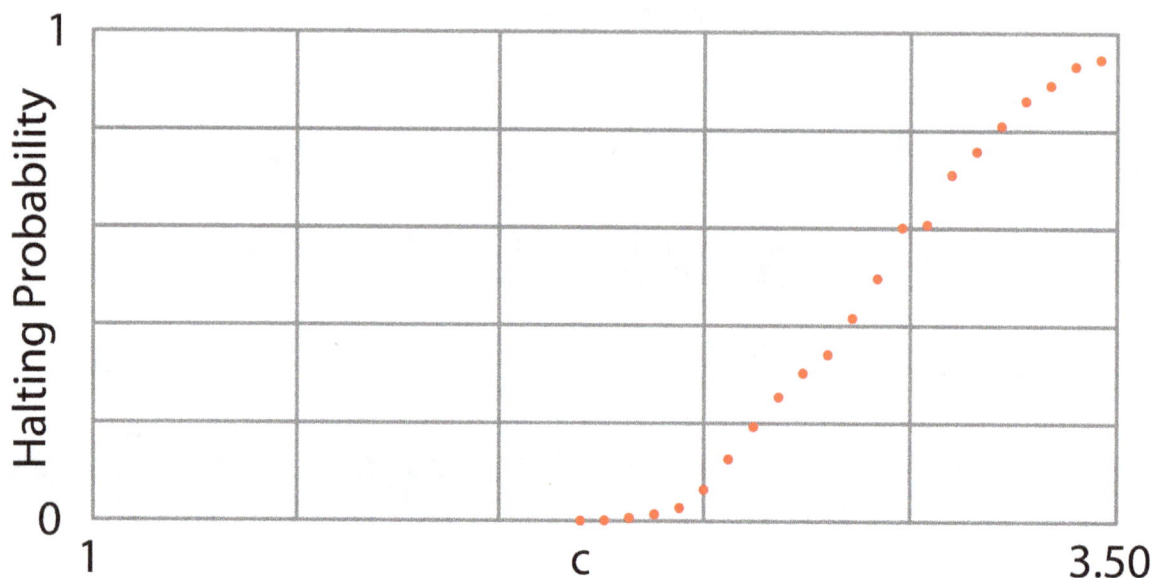

Fig. 9.2.   Halting probability for one-dimensional fractals using single segments with $N = 1$ (Inclusive boundary).

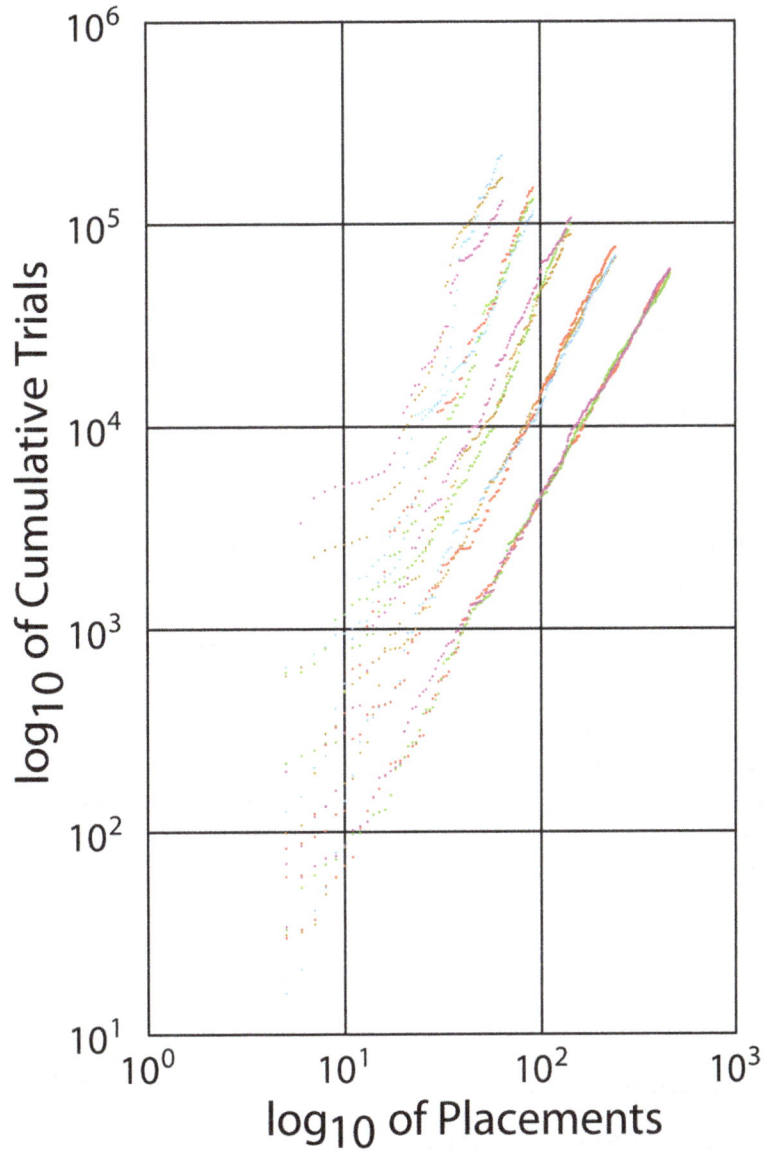

Fig. 9.3.   A plot of $\log_{10}(n_{\mathrm{com}})$ (vertical) for the cumulative number of trials versus $\log_{10}(n)$ (horizontal) where n is the number of placed segments. The grid lines are at powers of 10. In all cases, $N = 1$. The c values (from the bottom) are 1.7, 1.9, 2.1, 2.3, and 2.5. Three runs are plotted for each c value.

9.4. Fractal Dimension

The fractal dimension is given by

$$D = \frac{1}{c} \tag{9.1}$$

This result agrees with D values found by the box-counting method. Equations (7.11)–(7.20) in Chapter 7 can be adapted to one dimension, leading to Eq. (9.1) in another way.

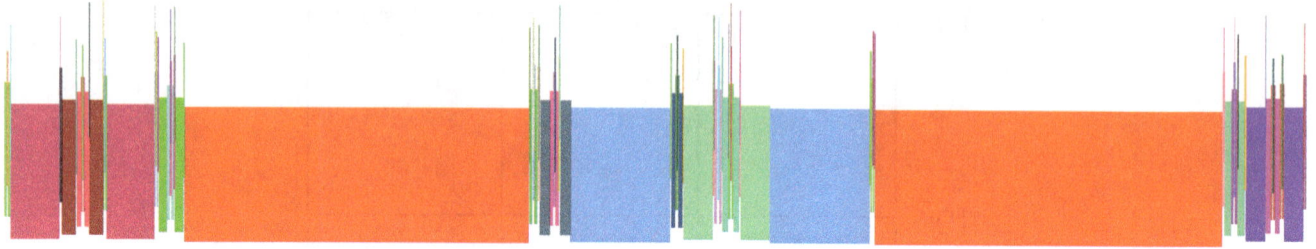

Fig. 9.4.   A bar graph of fractalized bisegments. $c = 1.8$, $N = 1$, 80 bisegments, 98% fill, inclusive boundaries, and random color. The two segments in each bisegment have the same color, and each segment starting with the longest is moved slightly upward for easier visualization. The finest features are unresolved.

9.5. Bisegments

One might think that it is not possible to have different shapes in one dimension, but this is not so. An example of a shape that can be defined is the *bisegment*. A bisegment here is defined as two line segments of width a, separated by a space of width a. It thus has a "hole" in the middle somewhat like the annular ring in two dimensions (Fig. 4.3 in Chapter 4).

Figure 9.4 shows a bisegment fractal close to the maximum c value. We can see the same type of nesting order as in Fig. 4.3 of Chapter 4 (annular rings).

For bisegments ($N = 1$) with inclusive boundaries, an upper limit on c is set by the fitting of the first few shapes. When $c > 1.7169$, there is no longer room for bisegment 1 at one end of the container segment even if bisegment 0 is at its other end, but there is room for bisegment 1 in the center of bisegment 0. The sum of the lengths of bisegments 0 and 2 is less than the container length until c reaches 1.9720 beyond which point no arrangement of the first three bisegments will fit. This configuration is seen in Fig. 9.4; bisegments 1 nests inside of bisegment 0, and bisegment 2 is between the end of bisegment 0 and the boundary of the container segment.

For single segments, it is not possible to construct a geometric argument for maximum c of the kind made in the previous paragraph.

9.6. Statistics and Dynamics of the Algorithm for Single Segments

The process begins by placing the first segment at random. This also creates two vacant intervals within the container segment, which can be filled during further placements. (If periodic boundaries are used, there is one vacant interval after the initial placement.) Since each new placement eliminates one interval and creates two new ones, there are $(n + 1)$ vacant intervals after placement of n segments. The database at any point in the process contains the locations of all placed segments and their lengths. It is thus possible to compute a complete set of vacant interval lengths after random placement n. When this is done, the intervals can be further sorted into those too narrow to accommodate the next-to-be-placed segment length (L_{n+1}), and those sufficiently wide. We will call the sufficiently wide intervals "gaps". Thus, at any point in the process, there is a number n_{gap} of available places to put the next-to-be-placed segment. This provides a sharp criterion for halting of the algorithm. *The algorithm halts if $n_{gap} = 0$* at any point. If $n_{gap} > 0$ for all n, the algorithm will continue to run and will eventually succeed for all n, even if an astronomical number of trials is needed.

The calculations described in this section were done with inclusive boundaries.

9.6.1. *The Distribution of Intervals*

The histogram in Fig. 9.5 was constructed by running the algorithm repeatedly with a fixed c, N, and 577 placed segments until data for 100 runs had accumulated. This gives good resolution. If continued "to infinity", this would

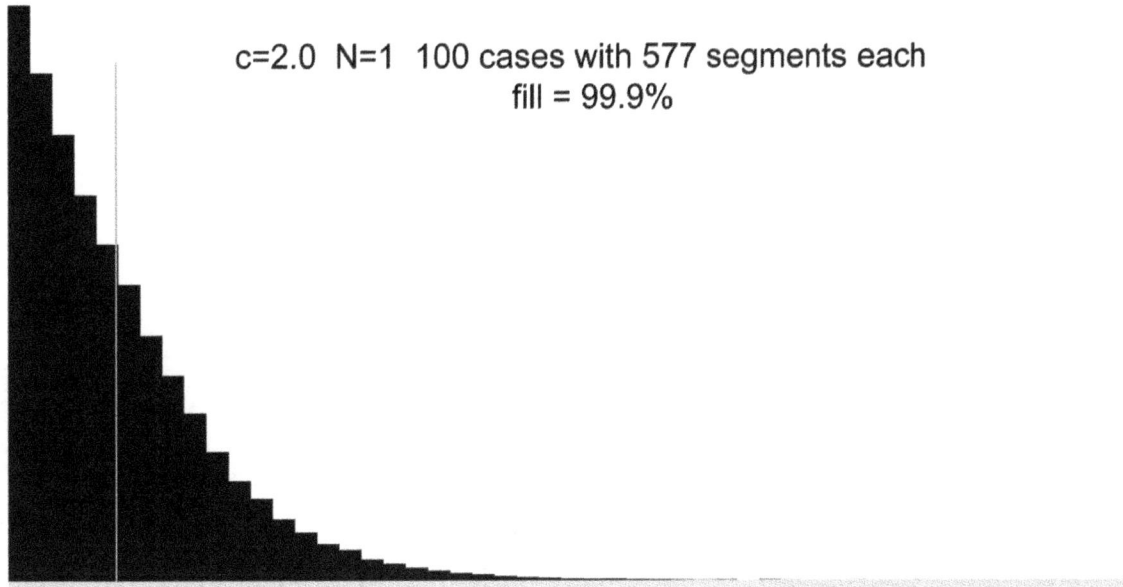

Fig. 9.5.   Histogram of the intervals between segments. $c = 2.0$, $N = 1$, 577 segments, and 99.9% fill. The vertical coordinate is the number of instances in 100 runs. The gray line underneath shows the extent of the data. The largest interval lies in the farthest-right bin. The vertical gray line shows the width of the last-placed segment on the same x scale.

define a continuous function which is the probability distribution function for the interval length. It can be seen that this is a falling exponential-like function with a large tail beyond the length of the next-to-be-placed segment.

There is a large population of intervals which are large enough to accommodate a new segment (histogram bars to the right of the vertical gray line). There is a sparse population of quite large gaps, which are improbably numerous if this is an exponential probability distribution function.

The "same at all scales" principle of fractals suggests that this distribution function, *when the x-axis is scaled to the value of the next-to-be-placed segment*, should be universal, i.e., the same at all n when n is large.

9.6.2.   *Trials and Placements*

We denote by $n_{cum}(n)$ the cumulative number of trials needed to place n segments. Because of the randomness, it is different for every run, but it has statistical trends. We define $n_t(n)$ to be the number of trials needed to make n placements. It has a very noisy behavior. Figure 9.6 plots these quantities on a log–log scale.

The process is quite noisy for small n, but as n increases the cumulative trials n_{cum} settles down to a power law which is reasonably well approximated by an exponent c (i.e., the black points make a line parallel to the green line). After the process has settled into a steady state, the log slope always agrees reasonably well with c, for all c and N values studied. The power law for n_{cum} with exponent c has been treated in Sec. 7.4 of Chapter 7 in two dimensions. There are places at about $\log_{10}(n) = 1.0$ and $\log_{10}(n) = 1.3$ where the black points go up very steeply, indicating that an unusually large number of trials were needed there. Such behavior is commonly seen in such records, especially for large c values where many trials are needed for each placement.

The n_t data is extremely noisy. Looking at the red points toward the right of the graph, we see that there are two points where it took more than a million trials to make a placement. Since the rules of statistics say that if the cumulative distribution follows a given functional form, the differential distribution is the derivative d/dn where we would expect the averaged n_t curve to have a slope $(c-1)$ (blue line) and this does not contradict the very scattered data. This data has $c = 2.6$ for which Fig. 9.2 gives a halting probability of ~0.2.

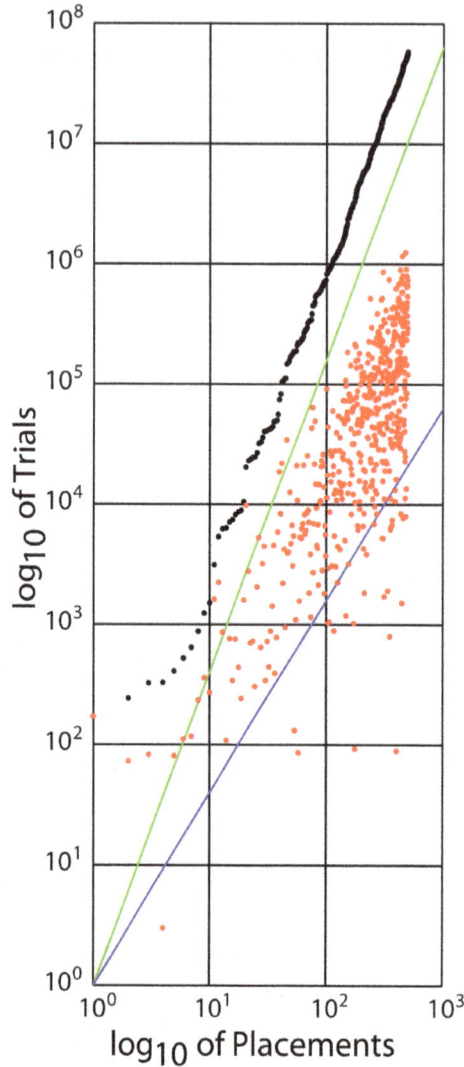

Fig. 9.6.   Trials. $c = 2.60$, $N = 1$, and 99.9977% fill. The black points are $\log_{10}(n_{com})$ versus $\log_{10}(n)$. The red points are $\log_{10}(n_t)$ versus $\log_{10}(n)$. The green line is a reference line with slope c, while the blue one is a reference line with slope $(c-1)$.

9.6.3. *Gaps*

As stated above, by a gap we mean an interval between placed segments which is large enough to hold the next-to-be-placed segment. The number of gaps (n_{gap}) will vary as the process continues. Figure 9.7 plots the number of gaps and the ratio gaps/$(n+1)$ versus placements n on a linear scale.

It can be seen that n_{gap} climbs steadily, but has some small up and down excursions. It appears quite unlikely that the red dot curve will ever dip down to zero (i.e., halting) for this particular run. There is mild oscillatory behavior, but it appears to be damped and not associated with instability.

The fraction of intervals which are gaps settles down to a steady 20–22%, which says that a probability distribution function (such as Fig. 9.5) for the probability p_{int} of finding an interval of width d, should obey the following scaling rule:

$$p_{int}(d) = f(d/L_{n+1}) \qquad (9.2)$$

The "same at all scales" principle says f should be a universal function (or at least independent of n for large n). The black points in Fig. 9.7 provide support for this idea.

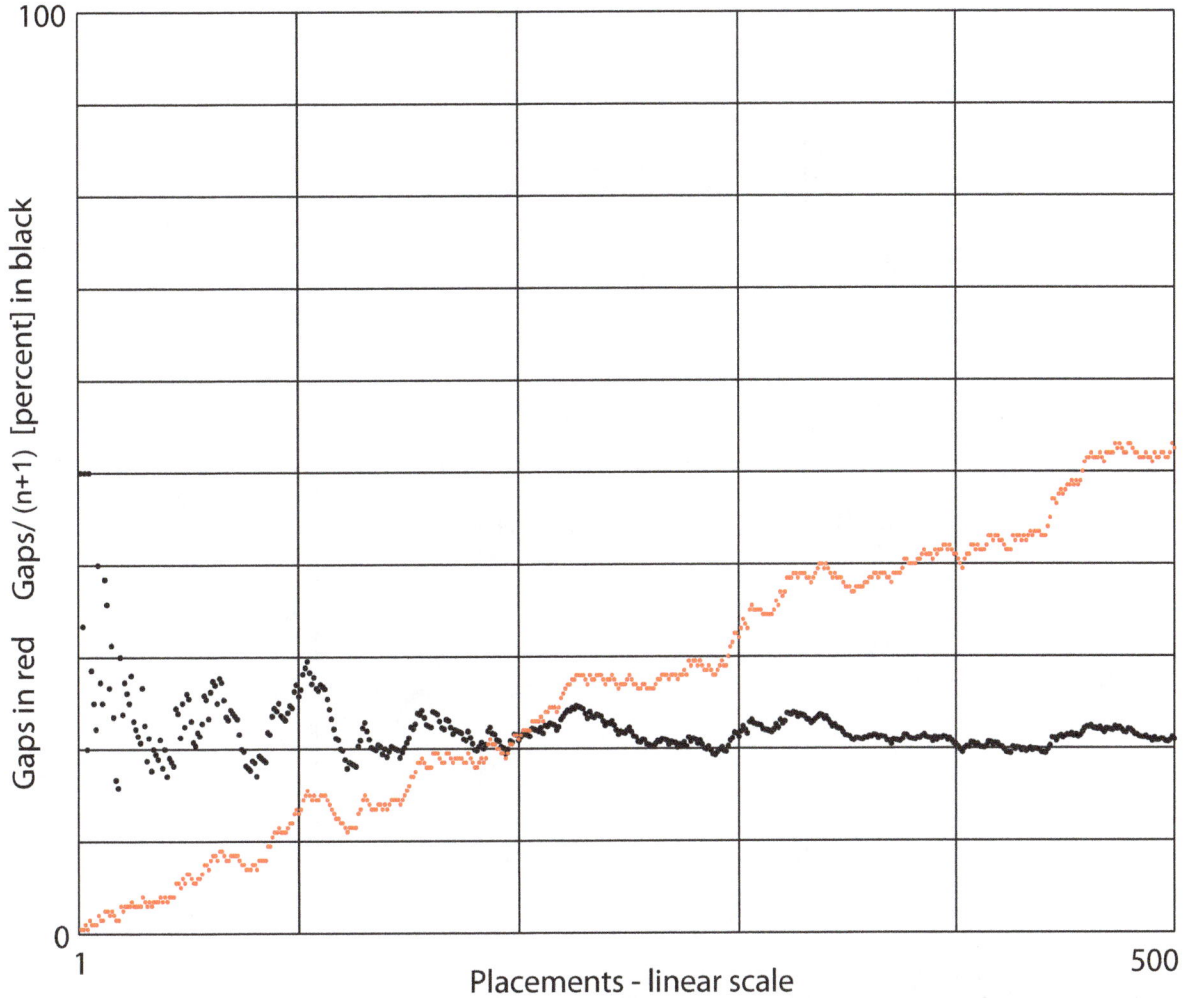

Fig. 9.7. Gaps. $c = 2.60$, $N = 1,500$ segments, and 99.9977% fill. The red points are n_{gap} values (integers). The black points are $n_{gap}/(n+1)$ as a percentage, i.e., the fraction of gaps relative to all intervals. The horizontal scale is placement number n, and is linear.

9.6.4. *The Placement Probability*

Since we have a complete list of the gaps, we can find the total amount of "free" (i.e., available for placement) length in all the gaps with any n. If we divide this by the total length L of the container segment, we find a placement probability P_{place}. This probability falls with increasing n (corresponding to more trials) and varies somewhat depending on the random numbers.

$$P_{place} = \frac{1}{L} \sum_{L_{gap} > L_{n+1}} (L_{gap} - L_{n+1})$$

(9.3)

We would expect the average number of trials for placement n to be proportional to $1/P_{place}$. We expect that the log–log plot of P_{place} should have a slope $-(c-1)$, the negative of that for the n_t data. It is seen that this fits well, and that the P_{place} data is far less noisy than the n_t data (red points in Fig. 9.6). Because the process is affected by

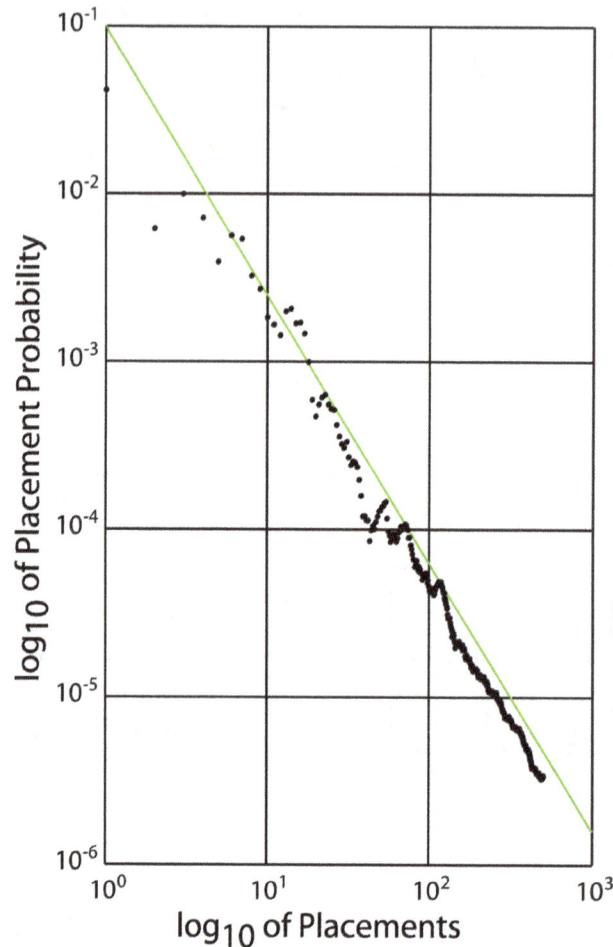

Fig. 9.8.   Placement probability. $c = 2.60$, $N = 1,500$ segments, and 99.9977% fill. The black points show $\log_{10}(p_{\text{place}})$ versus $\log_{10}(n)$. The green line has a reference slope of $-(c-1)$.

the particular random numbers used, there are dips and excursions in the data, but the general trend is a straight line. The dips, etc. in Fig. 9.8 show that this is a random process having memory. The very low values of p_{place} for large n agree with the huge numbers of trials needed for a placement with large n.

Since there are typically several gaps available to be filled at each stage, there is a type of feedback mechanism involved. If by random chance the largest gap is filled, the algorithm will have a harder time and take more trials at the next placement. On the other hand, if the smallest gap is filled, the algorithm will have an easier time at the next placement. Study of the run records shows behavior of this kind: a placement which takes an unusually large number of trials is followed by ones which take fewer. This may account for the dips and excursions of the data in Fig. 9.8.

9.7. The Dimensionless Average Gasket Width and the Halting Problem

For the one-dimensional case (assuming inclusive boundaries) the dimensionless average gasket width $\widehat{W}_{\text{ave}}^{\text{gasket}}$ after n placements is defined by

$$\widehat{W}_{\text{ave}}^{\text{gasket}} = \frac{L - \sum_{i=1}^{n} L_i}{(n+1)L_{n+1}} \tag{9.4a}$$

Table 9.1. Numerical values of the one-dimension dimensionless average gasket width versus placement number n with $N = 1$ for several c values.

	$N = 1$		
$\widehat{W}_{\text{ave}}^{\text{gasket}}$	$c = 1.6$	$c = 1.8$	$c = 2.0$
n =10	1.7132	1.2966	1.04683
n =100	1.6716	1.2549	1.00496
n =1000	1.6671	1.2504	1.00050
n =10000	1.6666	1.2498	1.00005
n =100000	1.6665	1.2490	1.00000

$$\widehat{W}_{\text{ave}}^{\text{gasket}} = \frac{\sum_{i=n+1}^{\infty} L_i}{(n+1)L_{n+1}} \tag{9.4b}$$

Here, L_i is the length of segment i. It is thus seen that the dimensionless average gasket width for this one-dimensional case is the ratio of the mean interval value (gasket width/(n+1)) to the next-to-be placed segment length L_{n+1}.

When $\widehat{W}_{\text{ave}}^{\text{gasket}} > 1$, the algorithm cannot halt[a]; there is always at least one gap greater than L_{n+1} which a sufficiently large number of trials will find. If $\widehat{W}_{\text{ave}}^{\text{gasket}} < 1$, halting is possible, but may not happen in an individual case, depending on the random numbers which occur. Table 9.1 shows computed values of $\widehat{W}_{\text{ave}}^{\text{gasket}}$ (with $N = 1$) versus n.

From its definition, $\widehat{W}_{\text{ave}}^{\text{gasket}}$ is independent of the random numbers used by the algorithm, and it can thus be computed by the usual methods. When $\widehat{W}_{\text{ave}}^{\text{gasket}}$ is calculated numerically, the results of Table 9.1 are found for $N = 1$. It can be seen that $\widehat{W}_{\text{ave}}^{\text{gasket}} > 1$ when $c < 2.0$ with $N = 1$. We see that $\widehat{W}_{\text{ave}}^{\text{gasket}}$ is a smoothly varying function of n. We can then conclude that *within this n range the algorithm is unconditionally non-halting* and thus space-filling. The nearly-flat behavior of this function versus n is similar to what was seen for $\widehat{W}_{\text{ave}}^{\text{gasket}}$ in the two-dimensional case.

The data of Table 9.1 cover the case of low n. But what happens for very large n? A rigorous asymptotic form or limit for $\widehat{W}_{\text{ave}}^{\text{gasket}}$ when $n \to \infty$ is a serious challenge, but a bound is somewhat easier. When $N = 1$, Eq. (9.4b) can be expressed as:

$$\widehat{W}_{\text{ave}}^{\text{gasket}} = \frac{\sum_{i=n+1}^{\infty} L_i}{(n+1)L_{n+1}} = \frac{1}{(n+1)}\left[\frac{\sum_{i=n+1}^{\infty} \frac{1}{i^c}}{\frac{1}{(n+1)^c}}\right] \tag{9.4c}$$

Comparison of the sum with the corresponding integral (convergent when $c > 1$) gives a lower bound for the sum:

$$\sum_{i=n+1}^{\infty} \frac{1}{i^c} > \int_{n+2}^{\infty} x^{-c}\,dx = \frac{(n+2)^{-c+1}}{(c-1)} \tag{9.5}$$

[a]The reasoning here is based upon the idea that if we have values $x_1, x_2, \ldots, x_n$ with mean value m, it must necessarily be true that at least one of the values x_j is greater than m.

thus for $N = 1$

$$\widehat{W}_{\text{ave}}^{\text{gasket}} > \frac{1}{(n+1)} \left[\frac{\frac{(n+2)^{-c+1}}{(c-1)}}{\frac{1}{(n+1)^c}} \right] = \frac{1}{c-1} \left[\frac{(n+1)^c}{(n+1)} \right] \left[\frac{(n+2)}{(n+2)^c} \right] = \frac{1}{c-1} \left(\frac{n+1}{n+2} \right)^{c-1} \tag{9.6}$$

finally

$$\lim_{n \to \infty} \widehat{W}_{\text{ave}}^{\text{gasket}} > \frac{1}{(c-1)} \tag{9.7}$$

It can be seen from Eq. (9.7) that when $c < 2$, $\widehat{W}_{\text{ave}}^{\text{gasket}} > 1$ in the limit $n \to \infty$. Thus, when $N = 1$ for $1 < c < 2$, the algorithm does not halt as $n \to \infty$. This result is in agreement with the numerical calculations of Table 9.1. The conclusion is that the one-dimensional algorithm is unconditionally non-halting with $N = 1$ if $1 < c < 2$. Thus, c_1 as defined in Fig. 7.2 of Chapter 7 should be 2, which is in general agreement with the numerical data in Fig. 9.2. A few small numerical studies for cases with $N > 1$ indicate that c_1 becomes larger as N increases.

It is concluded that there is a wide range of c values for which the algorithm is unconditionally non-halting for all n in one dimension and therefore it is space-filling in the limit.

9.8. Conclusions

The algorithm runs smoothly in one Euclidean dimension. The one-dimensional algorithm with single segments and $N = 1$ does not halt when $1 < c < 2$.

While it does not lead to the eye-catching images that can be made with the two-dimensional algorithm, statistical geometry in one dimension is easier to treat mathematically. The one-dimensional case may be the best place to start for serious mathematical studies of this algorithm. Most of the problems and conjectures listed for two dimensions in Chapter 7 also apply here.

Chapter 10

Do It Yourself

The algorithm is simple and easily stated. Creating your own code for the simple cases of squares or discs isn't rocket science, but without a reasonably good understanding of how the algorithm works you won't make much progress beyond that. You should study Chapter 7 or the Shier–Bourke paper in Computer Graphics Forum [8].

You will need to choose a graphics method. For those with good math skills but little exposure to graphics techniques, the use of "vector graphics" is desirable. In vector graphics, you only need to compute coordinates and dimensions of the shapes of interest. A disc, for example, can be described by the x and y coordinates of its center, and its radius r. The vector graphics statements must be interpreted in order to create an image in .jpg, .tif, or other popular raster graphics formats. This is done by OPENing the file, at which time the software will ask for the width and height of the image, and the number of pixels per inch. Based on this information the software will then "rasterize" the vector graphics data to produce the raster graphics image. For a red disc, the software will determine all of the pixel locations within the x, y, r region defined and color them red. Three numbers x, y, r may thus determine the content of a huge number of disc pixel locations. There are hundreds of ways to create vector graphics files.

10.1. The Hurwitz Zeta Function

The use of the Hurwitz zeta function may seem like a serious obstacle. It isn't on any pocket calculator. The routine below is written in C. (There are said to be more than 4000 coding languages in current use. C is about as close as we can get to a standard, but many readers will have a different system.) The following zeta routine may seem like "brute force" in its approach, but it has been tested against zeta function values from Mathematica and has at least six-decimal accuracy. In the code examples, the non-executable orange text is a comment.

```c
double zeta_hurw3(float u,float N)
{
  float nv;
  double v,sum,q,f;
  int Ni;
  long j;
  if(N<.4) printf("WARNING: untested for accuracy with N < .4\n");
    Ni=N;    // find integer part
  f=N-Ni;  // excess over next-lowest integer
  sum=0.;
  j=Ni;  // start at Ni
```

```
    do   // sum a huge number of terms
    {
        nv=f+(float)j;
        q=pow(nv,-u);
        sum=sum+q;
        j++;
    }while(q>.0000001);
    v=sum+est_sum3(u,(double)j+f);   // estimate the rest of the sum
    return v;
}


  double est_sum3(float c,double n)
{
    double u,v;
    u=(double)n+.5;
    v=(1./(c-1.))*pow(u,1.-c);
    return v;
}
```

10.2. Fractalization of Squares

This code is also written in C. Note the call to the Hurwitz code. It is hoped that with enough comments the user can adapt this to a wide variety of code languages. The code has only 27 executable statements.

```
// code for fractalizing squares inclusively in a rectangular boundary
// wdpix and htpix define the image width and height
void place_squares_inclusive(float cval,float Nval,int shapesmax,
float wdpix,float htpix)
    {
        float xb[4096],yb[4096],halfside[4096]; // storage for placed shapes
        float area1,x1,y1,s2,dx,dy,r12,q1;
        float zval,acoeff,afac;
        int tst1,j,nshape,k;
        afac=4.;                    // area for half-side-length = 1
        zval=zeta_hurw3(cval,Nval);    //compute zeta function
        acoeff=htpix*wdpix/zval;            //factor for size calculation
        //  ------- start of computational loops ---------
        nshape=0;    // placement number initialization
        do   // loop on placements
        {
            q1=(float)nshape+Nval;
            area1=acoeff*pow(q1,-cval);   // trial area
            s2=sqrt(area1/afac);        //trial half-side
            do     // loop on random search
            {
                tst1=1;    //initialize test
                //the function rd() creates a random number between 0 and 1
                x1=s2+rd()*(wdpix-2*s2);    //x,y for random trial
```

```
        y1=s2+rd()*(htpix-2*s2);
        for(k=0;k<nshape;k++)         // loop over old placements
        {
            r12=(halfside[k]+s2);
            dx=fabs(xb[k]-x1);
            if(dx<=r12)     //early quit if delta_x too large
            {
                dy=fabs(yb[k]-y1);
                if(dy<=r12)   //early quit if delta_y too large
                {
                    tst1=0;    //there is an overlap
                    break;     //exit the for loop
                } // if(dy
            } // if(dx
        } // next k
    }while(tst1==0); // repeat search if overlapping a prior shape
    printf("%i  %f  %f  %f  \n",nshape,x1,y1,s2);  //optional
    xb[nshape]=x1;    //put successful placement data in the data base
    yb[nshape]=y1;
    halfside[nshape]=s2;
    nshape++;            //next square -- increment nshape
  }while(nshape<=shapesmax);  //end of loop on nshape
//code below creates a text file with the placement data
// the file with identifier cd must be OPENed in another part of the code
  fprintf(cd,"w %f h %f %ld \n",wdpix,htpix,shapesmax);
  for(j=0;j<=shapesmax;j++)    // save data to file
    fprintf(cd,"%9.5f   %9.5f   %9.5f \n",xb[j],yb[j],halfside[j]);
}
```

This code doesn't create a graphics file. It simply creates a database of the sizes and locations of the placed squares. The code for **rd()** is special to the author's system. It creates a random number between 0 and 1. Many coding languages have functions for creating a random number and if one is available it should be used.

This code illustrates some general features. The system of **while** and **for** loops seen here is used in the code for all of the examples in the book. The "early quit" statements basically say that if the two squares are impossibly far apart, they cannot possibly overlap and one can quit now. For small squares, this often terminates the test at an early stage saving computation time. If they are far apart in x, no y test is needed. For more complicated shapes, one can define an "envelope" rectangle which lies entirely outside the shape and do the early-quit test on that. This minimizes the number of detailed overlap tests needed, since the detail test is only performed if the envelope rectangles overlap each other.

The integer **tst1** is used as a TRUE–FALSE variable. There are doubtless more elegant ways of doing this.

Because of the huge variety of methods for graphics, the creation of an image is left to the reader. The author's work uses generic Postscript, and his documentation is "Postscript by Example" by McGilton and Campione. It is doubtless out of print, but used copies may be available. Generic Postscript files have the nice feature that they are human-readable. With this graphics approach, a vector-graphics .ps file can be created with the **fprintf** statements in C, or equivalent statements in other languages. Such files can be OPENed with Photoshop and many other programs, and used to generate bit graphics files in .jpg and other formats. The program rasterizes the vector graphics regions, coloring all the points in them with the specified color.

10.3. Overlap Tests

Squares and rectangles are the simplest tests. If we are dealing with discs, the innermost **if** test in Sec. 10.2 is replaced by the following statements:

```
if(dy<=r12)   //early quit if delta_y too large
{
  u=pow(dx,2)+pow(dy,2);    //sum of squares
  ru=sqrt(u);      //ru is the diagonal distance between the disc centers
  if(ru<r[k]+rtest)   //is ru < sum of the radii of the discs?
  {
      tst1=0;    //if true there is an overlap
      break;     //exit the for loop
  }
} // if(dy
```

More complicated disc arrangements (e.g., Figs. 4.7 and 4.9 in Chapter 4) can be overlap tested by testing overlap of all discs in shape A against all discs in shape B with quit on first fail. The same thing can be done with rectangles as illustrated in Fig. 2.10 of Chapter 2.

We can divide overlap tests into two kinds: In exact overlap tests, the only limitation on accuracy is the resolution of the floating-point numbers used. The tests for squares and discs are of this kind. There are also approximate overlap tests, and many of them have been used in this work. Some software systems such as Matlab have routines for testing the overlap of two shapes, with the shapes usually defined as polygons with an arbitrary number of sides.

Discs and rectangles have simple exact overlap tests. Overlap testing for annular rings (Fig. 4.3 in Chapter 4) is an exact test. Simple polygons often have exact overlap tests.

Most of the author's approximate overlap tests, are based on the idea that if you have a large number of test points defined along the edges of shape A, and none of them lies within shape B, there is no overlap. One can only get an exact test this way by using an infinite number of test points, but you can get quite good results with a finite number of them. In some cases, I have used up to 50 or 60 test points. This requires you to define two things: a set of test points and a test for whether a point x, y lies within a particular fill region. A point test is much simpler than a full region-overlap-region test. We turn to a particular case: a butterfly.

Figure 10.1 illustrates the steps in setting up the butterfly. The first task was to define the butterfly shape. Local polar coordinates were used. The wing shape was developed by trial and error, using a C function which draws the butterfly to evaluate success. The next stage was development of code testing whether a point x, y lies within a butterfly. This was tested by creating a large number of random x, y points and only drawing them if the test says they fall inside the butterfly. These are the sprinkled dark dots which show that the test works. The last stage is to set up the test points at the edge of a butterfly, and these are shown as red dots (there are 36 of them). A drawing program to "test the tests" as in Fig. 10.1 is very helpful for avoiding errors.

If your first fractalization run (e.g., Fig. 10.2) shows overlaps, the fix is to go back to the "test the tests" program of Fig. 10.1 and add more test points.

It is a rather general truth that one must place test points at convex regions of the shape, but few or none in concave regions. The sharper the curvature, the more points are needed there.

The butterfly of Fig. 10.1 is rather generic. The final task in the development was to devise colors and patterns which look like a real butterfly. The iconic North American "monarch" butterfly was chosen. There are enough differences between male monarchs and females that one must choose. This one is a female. The pattern and colors required a lot of tedious trial and error, but the result (Fig. 10.2) is quite pleasing. We know that the overlap test is not exact here, but we don't see anything that looks like an overlap in the finished image. It would be quite difficult to include an overlap test for the antennae, so they are allowed to overlap other butterflies. This butterfly image was entered in the math-art show at the 2017 JMM conference, and won Best of Show!

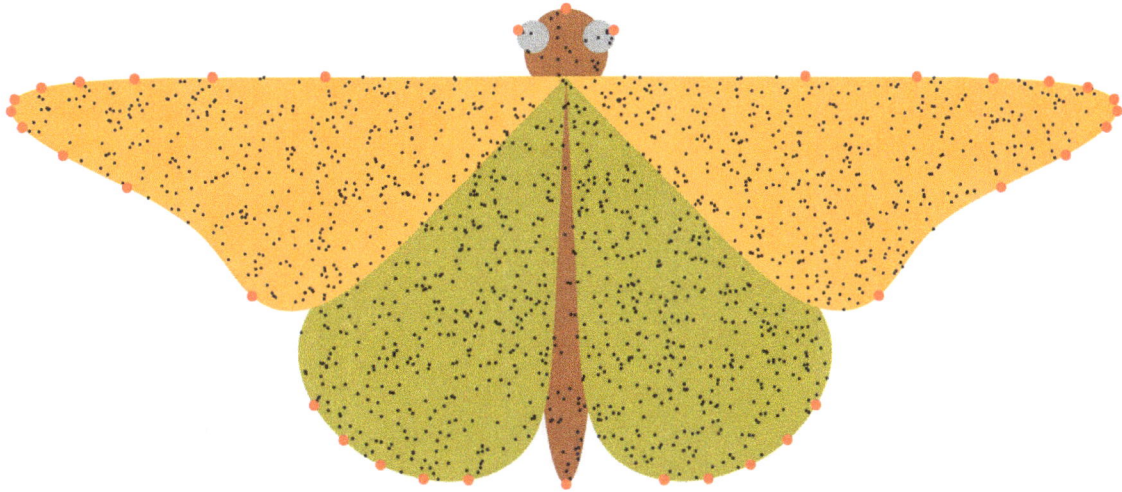

Fig. 10.1. A butterfly fill region with setup for an overlap test where edge points (red dots) in butterfly A are tested for inclusion in butterfly B.

Fig. 10.2. Art and mathematics. Fractalization of monarch butterflies.

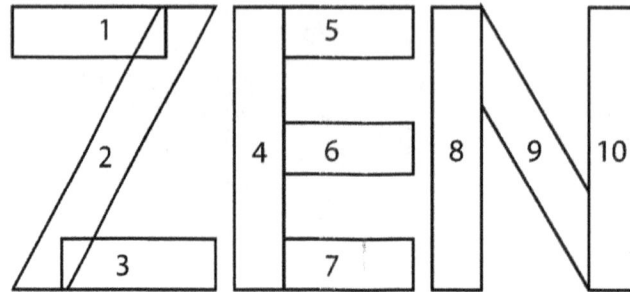

Fig. 10.3.   An example of a fill region built up from several smaller ones.

Development of an overlap test can begin with the creation of a function of the following form:

```
void draw_the_shape(float x, float y, float h)
```

Here, x and y are the coordinates of the shape relative to its origin of coordinates, and h is its *scale factor*. In physics terms, the scale factor is a length (units meter). This function draws the fill region shape. The most obvious choices for a disc are the x and y coordinates originating at its center, and its radius. But such a choice is not unique. You could just as well put the origin of coordinates at the left boundary and use the diameter as the scale factor if you set things up suitably. The scale factor is somewhat arbitrary — for the butterfly, it could be the vertical head to tail length, or the wingspan. Once you choose it, you must use it consistently.

It is often desirable to build up a fill region from a number of smaller regions. This is illustrated in Fig. 10.3 with the word ZEN made from ten quadrilaterals. The test for "x, y in the fill shape" can be created using x, y -in-shape tests for each of the individual regions. One starts with a particular region and cycles through all of them, quitting on first fail, i.e., if you find a test point within one of the subregions, there is overlap and you go on to the next trial. The average time needed for the test is influenced by the order in which you do the tests. It is best to start with those regions farthest from the center — regions 1, 2, and 10 here. This is the method used in many of the examples. The regions can overlap, as seen with 1, 2, and 3 in Fig. 10.3.

10.4. Fill Region Area

You also need to know the areas of the butterflies, etc. In the butterfly example, there is no textbook formula for it. Set up the "x, y within the fill region" test with unit scale factor. Create a dense rectangular grid of x, y points that extends beyond the butterfly and test every point for in or out. Add up all the in points and multiply by the unit cell area of the rectangular grid, and you have an approximate calculation of the area. Tests with variable number of points show that if you have 5–10 million points inside the fill region the accuracy of the area calculation is about 1 part in 10^5. This may be slow computation, but it only needs to be done once. This accuracy is good enough for graphics work. If the area with unit scale factor is A', the area with any other scale factor is $A = A'h^2$.

10.5. Running Your Code

It is easy to stumble over the choice of parameters c and N for your first test run. With the wrong choice, the algorithm will halt (see Fig. 7.3 in Chapter 7).

Choose a low (but greater than 1) value for c. A low value ensures that the algorithm does not halt. Choose a "high" value for N. A high value ensures that the first shape will fit within the bounding region. For discs or squares, c = 1.25 and N = 2 is a reasonable choice for starting. Sparse or sprawly shapes will need lower c values and higher N. c = 1.1 almost always works.

If the algorithm runs with your initial choice of c and N, you can increase c and/or decrease N to produce the kind of image you want. An important advantage of low c for your first runs is that few trials are needed and the computation is fast. As you increase c, it takes more trials to find a placement and the process slows greatly. Slow computation with high c values sets a practical upper limit on c.

10.6. Area–Perimeter Calculations

As an example, we give code which repeats the calculations for the area–perimeter algorithm for the case of three-dimensional spheres.

```
void vol_area_calc(float gamma,long shapesmax)
  {
  double r_k,vol1,Vgask,Agask;
  double pii,fourthirds_pi,fourpi;
  float q1,q2,q3,logVg,logi,logV,dlgV,vval[32],logslope;
  int i,imin,imax;
  long k,kt,n;

  pii=4.*atan(1.);         //compute pi
  fourthirds_pi=(4./3.)*pii;
  fourpi=4.*pii;

  Vgask=fourthirds_pi;     //initial values -- bounding sphere of radius 1
  Agask=fourpi;
  i=2;
  imin=i;
  kt=2*i;
  for(k=1;k<=shapesmax;k++) // recursive fill region volumes
  {
      r_k=gamma*Vgask/Agask;            //next radius
      vol1=fourthirds_pi*pow(r_k,3);    //fill sphere volume for next radius
      Vgask=Vgask-vol1;                 //decrement gasket volume
      Agask=Agask+fourpi*pow(r_k,2);    //increment gasket surface area
      if(k==kt)  //save log fill sphere volume at power of 2
      {
        vval[i]=log10(vol1);
        kt=kt*2;
        i++;
      }
  }
  imax=i-1;

  printf("    iteration       log10(Vf)    logslope\n"); //column heads
  for(i=imin;i<=imax;i++)       //print the results
  {
      if(i==imin) dlgV=0.;
      else dlgV=vval[i]-vval[i-1];
      logslope=dlgV/log10(2.);
      n=pow(2,i);
      printf("     %9i    %10.6f  %10.6f \n",n,vval[i],logslope);
      if(i==imax)
      {
```

```
      q2=dlgV/log10(2.);
      printf("computed sphere volume exponent c %f \n",logslope);
      }
   }
q3=(9.+3*gamma)/(9.+2.*gamma);   //c from formulas of Chap. 8
printf("gamma %f   c from formula %f\n",gamma,q3);
}
```

Note the use of double-precision variables. Because of the very lengthy chain of computations, with each iteration depending on the one before, roundoff error is a concern. Independent checks showed little roundoff error with the software used.

The tabulation below shows the output. The first column is the iteration number (placement number). The second column is $\log_{10}$ of the placed sphere volume, and the third column is the difference between the second column entry and the one above it in the list, divided by $\log_{10}(2)$. The computed value agrees to almost six decimal places with the value found from (Chapter 8):

$$c = \frac{9+3\gamma}{9+2\gamma}$$

The computed results are:

iteration	log10(Vf)	logslope
4	−1.221093	0.000000
8	−1.479679	−0.859004
16	−1.763508	−0.942859
32	−2.064780	−1.000804
64	−2.377256	−1.038023
128	−2.696553	−1.060683
256	−3.019849	−1.073966
512	−3.345427	−1.081547
1024	−3.672283	−1.085791
2048	−3.999844	−1.088134
4096	−4.327790	−1.089414
8192	−4.655946	−1.090109
16384	−4.984213	−1.090481
32768	−5.312541	−1.090683
65536	−5.640902	−1.090790
131072	−5.969279	−1.090846
262144	−6.297666	−1.090876
524288	−6.626057	−1.090891
1048576	−6.954450	−1.090899
2097152	−7.282845	−1.090904
4194304	−7.611241	−1.090907

```
computed sphere volume exponent c       −1.090907
gamma 1.000000     c from formula −1.090909
```

The tabulation above also gives some idea how fast the calculation of c is converging with more iterations. This is in excellent numerical agreement with a c equation containing integer coefficients. The reader is reminded that this computation is purely about numbers, unaffected by random trials.

References

[1] I. Stewart, *Symmetry: A Very Short Introduction* (Oxford University Press, Oxford, 2013).

[2] F. Bool, J. Kist, J. Locher and F. Wierda, *M. C. Escher: His Life and Complete Graphic Work* (Harry N. Abrams, 1981).

[3] B. Mandelbrot, *Fractals: Form, Chance, and Dimension* (W. H. Freeman, San Francisco, 1977).

[4] P. S. Dodds and J. S. Weitz, "Packing-limited growth," *Phys. Rev. E*, **65**, (2002) 056108.

[5] P. S. Dodds and J. S. Weitz, "Packing-limited growth of irregular objects," *Phys. Rev. E*, **67**, (2003) 016117.

[6] G. W. Delaney, S. Hutzler and T. Aste, "Relation between grain shape and fractal properties in random apollonian packing," *Phys. Rev. Lett.*, **101**, (2008) 120602.

[7] J. Shier, "Filling space with random fractal non-overlapping simple shapes," in *Proceedings of the ISAMA-11 Conference*, Chicago, IL (2011) 131.

[8] J. Shier and P. Bourke, "An algorithm for random fractal filling of space," *Comput. Graph. Forum*, **32**(8), (2013) 89–97.

[9] D. Dunham and J. Shier, "The art of random fractals," in *Proceedings of the 2014 Bridges Conference*, Seoul, Korea, 79–86.

[10] D. Dunham and J. Shier, "Fractal wallpaper patterns," in *Proceedings of the 2015 Bridges Conference*, Baltimore, MD, 183–190.

[11] D. Dunham and J. Shier, "Repeating fractal patterns with 4-fold symmetry," in *Proceedings of the 2016 Bridges Conference*, Jyvaskyla, Finland, 523–524.

[12] C. Ennis, "(Always) room for one more", *Math Horizons*, February issue, (2016) 8.

[13] D. Dunham and J. Shier, "Artistic patterns — from randomness to symmetry," *Symmetry: Culture and Sci.*, **27**(4) (2016) 257–448.

[14] M. Buchanan, *Ubiquity* (Three Rivers Press, New York, 2000).

[15] P. Ball, *Critical Mass* (Farrar, Straus and Giroux, New York, 2004).

Index